SUSTAINABLE DEVELOPMENT :
Issues and Considerations

SUSTAINABLE DEVELOPMENT :
Issues and Considerations

PRATYUSH MANI TRIPATHI

SERIALS PUBLICATIONS
NEW DELHI

First Published - 2005

ISBN : 81-86771-91-3

Published by

SERIALS PUBLICATIONS

4830/24, Prahlad Street, Ansari Road
Darya Ganj, New Delhi-110002 (India)
Phone : 23245225. Fax : 91-11-23272135
E-mail: serials@satyam.net.in

Preface

The concept of sustainable development is of recent origin and has gained popularity because it is closely linked with existence of life on the planet earth. Throughout the globe, ecological imbalances are growing because of heavy population growth and poverty and India is no exception. The quality of Indian environment is deteriorating at fast pace. Aggressive exploitation of the environment started during the colonial rule, gained momentum in the post independence era and today, it has assumed alarming proportions. It is estimated that the cost of ecological degeneration in India is more than 6% of her GDP. Hence, there is a need to sensitize masses about the scale of environmental degeneration and its causes.

In this context, sustainable development has become a very relevant issue for research. The present study deals with the threats and challenges of development in India in general and eastern U.P. in particular. For the intensive study I have taken Gorakhpur district as the field area. Rich in land and other resources, eastern U.P has been primarily an agrarian society. Even after five decades of planned development programme, it is still considered to be a backward region of the country. The problem deeply lies in the type of developmental model which has a wrong set of priorities. It has not paid sufficient consideration to the optimum utilization of natural and human resources. Had it been properly harnessed then migration from rural to urban have been controlled, thus controlling or minimizing congestion and growth of slums in the towns, rehabilitation and marginalization of the indigenous people.

In the present study, I have reviewed various theoretical issues and values of sustainable development. The study tries to visualize the problems of sustainable development through the case study of Gorakhpur district. It has also included the factors of population, poverty, political dimensions and people's participation. Special emphasis has been given on the factor of human development.

In the completion of this work, I have been wholeheartedly supported and helped by many. Prof. K. Kishore of Public Administration Department of Lucknow University, supervised the whole work enthusiastically and corrected me from time to time. Dr. C.P. Barthwal, professor and head, Department of Public Administration, Lucknow University, Lucknow made me learn many conceptual issues. Dr. B. W. Pandey of Department of Geography, Sahid Bhaghat Singh College, University of Delhi provided a lot of materials, Dr. S. K. Chaudhary of Department of Sociology University of Lucknow, went through the first draft. I am greatfull to all of them. I must pay my gratitude to my family members for their unstinted psychological support for the completion of this work. I cannot forget to mention my thanks to Sri Ritesh Kumar Saxena who typed and provided time whenever I asked him to do so.

Finally I thank Sri S.K. Jha of Serials Publications for taking active interests in the publication of this book.

Pratyush Mani Tripathi

Place: Lucknow

Dated: Februray, 2005

Foreword

The development debate, specially in the developing countries, criticizes the popular catchword and calls for new paradigms of development, not only because the current ones do not seem to be working well, but also because the university of western paradigms of development is proving to be nightmare and has originated as a reaction to the ascendancy of the economic/capitalist model of development. This research study had discerned cracks in the very notion of development which the capitalist FIRST WORLD prescribed as a panacea for the post-colonial THIRD WORLD. This research study on alternative development questions the dominant position of the economist paradigms of development models which envision the primacy of economic dimensions of development over the non-economic ones.

For us, sustainable development is both a challenge and an opportunity. The obstacles are, however, great and making the concepts of sustainability precise, is difficult. It is not possible to argue that there should be zero use of natural resources for development. Successful development will inevitably involves some amount of depletion of natural resources, resulting in environmental damage. Future policies and programmes of accelerating environmentally responsible development will not happen by themselves. It is, therefore, important to seize the current opportunity to bring about real, if not radical change.

The idea of sustainable development, though not entirely new and hitherto unknown, was introduced by Brudtland Commission, 1987, in its Seminal Report "Our Common Future" and popularized by World Bank and United Nations Environment Programme. As such "sustainable development is development that meets the needs of the present without compromising the ability of future generations to meet their own needs". The *Earth Summit* at Rio-de-Generio (1992) and Johnannesburg city on 2nd December 2003 luckily came of with two alternative vision of development under the wider concepts of sustainable development which provide centrality to non-economic dimension of development and effectively challenge the supremacy of economic growth model that held sway for long. These two alternative visions of development revolutionized the development thinking across the world and more so in the developing countries which were already getting disenchanted with pursuing a materialist dream.

This book analyses the concept sustainable development arising out of environment-development debate. It discusses the strategies of sustainable development in the context of Gorkhpur District arguing that transparent environmental and economic policies backed up with a strong political will is needed to realize the goals of sustainable development.

It focuses on a wide variety of political issues, such as democracy, social integration, human rights, human security, globalization, structural adjustment, etc. which inevitably enter into any informed research on development. The book exposes the rhetoric and reality of participatory development as its hallmark. It cites examples of how development programme based upon people's participation paradigm lost their momentum due to deep-seated aversion of decision makers and powerbrokers to meaningfully involve the people (the beneficiaries) in the conceptualization, planning, implementation, monitoring and management of development projects.

The author has seen the functioning of development machinery with all possibilities and threats. The chapters are meaningfully arranged and classified. No doubt the book will make an interesting reading and thoughtful source of information for the general readers and researcher alike.

Lucknow
Dated : 05-05-05

Dr. C.P. Barthwal
Professor & Head
Department of Public Administration & Dean,
Faculty of Arts University of Lucknow

Contents

1
Introduction

Sustainable Development : Issues and Considerations

Sustainable development is one aspect of the larger issue of development of human society today. The concept of development has become a catchword in the 20th century, particularly in the later half. Initially it was the economist's prerogative, who mostly concentrated on economic indices such as growth, per capita income, gross national product etc. without considering much the issue of redistribution. Later on, the concept of social development emerged in the 1970s, which took the aspect of change of social structure, bringing in modernity, rapid industrialization, advancement of generation, equitable distribution, women development and bridging the gap between rich and poor amongst other factors. Thus, in the 1970s and 1980s, modernisation, development and industrialization became almost synonymous. It had many side-effects, in the sense that it led to maddening consumerism, environmental degradation and plundering of resources without any perspective planning, hence came the concept of sustainable development which envisages development with conservation.

The term sustainability originates from the Latin word *sustinere*, which means 'to hold up', 'to endure'. *The Oxford Reference Dictionary* (1980) refers sustain as 'endure without giving way', 'to support especially for a long period'. The meaning given by *Collins Dictionary* (1979) is 'to maintain', "to support'. Similarly, *Webster's New Collegiate Dictionary* (1975) describes sustainability as 'to give support', 'to keep up'; 'prolong'. From these, one can infer that the general meaning of sustainability is to maintain or to support a programme/ project for a long period, which is simply not possible without effective and efficient participation of people and communities.

Sustainable development is defined by UNDP in its *Human Development Report (1991)* as "development that improves health care, education and social well-being". Such human development is now recognized as critical to economic development and to early stabilization of population. It further states that "men, women and children must be the center of attention with development woven around people and not people around development".[1]

Goodland and Ledec (1987) stress that, "using renewable natural resources in a manner that does not eliminate them or otherwise diminish their 'renewable usefulness' for future generation as sustainable development".[2]

World Conservation Union (1991) defines sustainable development as "improving the quality of human life while living within the carrying capacity of supporting eco-system".[3] The Report focuses on sustainable development as a process requiring simultaneous global progress in a variety of dimensions: economic, human, environmental and technological.

Sustainable development requires meeting the basic needs of all and providing opportunity to satisfy their aspirations for a better future life. The concept goes beyond the parameter of self-reliance and self-sufficiency though the two are often attained together. Sufficiency does not necessarily always ensure sustainability. In short, sustainable development calls for multifaceted and comprehensive interventions to address wide varieties of problems, simultaneously using different types of policy approaches, each designed specifically with reference to a particular issue. Resilience is the watch-word of this concept.

The term "sustainable development" was brought into common use by the World Commission on Environment and Development (Brundtland Commission) in its seminal 1987 report, "Our Common Future". The idea of sustaining the earth has proved a powerful metaphor in raising public awareness and focussing on the need for better environment stewardship.

The Brundtland Commission's definition of the term, "meeting the needs of the present generation without compromising the needs of future generations", is strongly endorsed by the Human Development Report (1991). We also believe, with the Brundtland Commission, that in meeting the needs of poor in this generation, there is no difference between the goals of development policy and appropriate environmental protection. Both must be designed to improve welfare.

The World Conservation Strategy (1980) defined sustainability as "the management of the human use of the biosphere so that it may yield the greatest sustainable development to present generation while maintaining the potential to meet the needs and aspirations of future generations".[4]

Making the concept of sustainability precise however, has proved to be difficult. It is not plausible to argue that all natural resources should be preserved. Successful development will inevitably involve some amount of land clearing, oil drilling, river damming, and swamp draining. Some have argued that natural capital should be preserved in some aggregate sense, with losses in one area replenished elsewhere. This approach has been helpful in focussing attention on the need to estimate the value of environmental resources and on the importance of protecting certain essential ecological system.

Sustaining Development

In terms of incomes and output, the world will be a much richer place later in this century. Here, several questions emerge—Will the environment be richer in the next century or will the environment be much poorer? Will future generations be worse off as a result of environmental degradation that results from economic decisions made today? Will increase in the scale of economic activity be sustainable in the face of increasing pressure on natural resources? Prospective changes of the size described above-raised fundamental questions about the kind of world we will bequeath to our children and about the nature and goals of development.

What is Development?

Development can be defined as a desired change which has three aspects—change orientation, goal orientation and progressive orientation. Development is about improving the well being of people. Raising living standards and improving education, health and equality of opportunity, are all essential components of economic development. Ensuring political and civil rights is a broader developmental goal. Economic growth is an essential means for enabling development, but in itself, it is a highly imperfect proxy for progress.

The first step in improving social choices is to measure progress correctly. It has long been recognized that measures of, for example, educational opportunity, infant mortality, and nutritional status are essential components of GDP or GNP. Some have even tried to merge these indices to capture progress in development index constructed by the United Nations Development Programme (UNDP) in such an effort. The fact that environmental damage hurts people, both today and in the future, provides additional ground for rethinking our measurement of progress. Indeed it raises special concerns for education, health, nutrition, and life expectancy, with a tendency to improve by economic growth and social development.

The environment is sometimes damaged by that growth. Furthermore, the people suffering from the damage are more likely different from those enjoying the benefits of growth. They may, for example, be today's poor, or they may be future generations who inherit a degraded environment. For these reasons it is essential to assess the costs of human welfare to environmental damage, a central theme of this Report –and to take account of the distributional impacts of policies, particularly for the poor.

As said earlier, the World Commission Report On Sustainable Development says, it is "development that meets the goals of the present without compromising the ability of future generations to meet their own needs" (WECD, 1987-1988). This pioneering definition implies a view of sustainable development as a case of inter-generational sensibility in respect of the use of natural resources. It contains two basic ideas; one, that of the right of the future generations, to inherit all ecological and natural resources which the present generation inherited from the preceding one; two, that of the responsibility of the present generation to use available ecological and natural resources in a way that these are passed on intact to the succeeding generations. The second idea holds the key to sustainable development as is evident from the following formulation of The World Food Programme: "To allow for future generations, require that we preserve our remaining resources and that we heal or rehabilitate resources that have been treated carelessly in the past. To do so these things systematically, is to follow a path of environmentally sustainable development". In concrete terms, sustainable development focuses on three things:

1. To follow a path of development that does not impair or damage the cover provided by the nature to mankind,

2. To use the renewable natural resources so that their rate of re-generation is always in excess of their rate of use, and,

3. To use the non-renewable resources in a sparing and responsible manner and to ceaselessly work to find substitutes for them.

Sustainable development is thus different from conventional development. Conventional development was so obsessed with growth that it gleefully neglected ecology; indeed, it acted detrimental to nature. Sustainable development, on the other hand, refers to ecological friendly development, one that harnesses nature without destroying it, keeps up its greenery and conserves its resources.

Originally defined within the context of ecological harmony, more recently sustainable development has taken on some broader interpretations. Apart from its ecological concern, sustainability is now being defined also in terms of technical, managerial, institutional, human, social and cultural dimensions. In the latest literature on the subject there has emerged the idea of "multiple sustainabilities"[5]. The expanding frontiers of sustainability notwithstanding, the original definition of sustainable development as the capacity to provide for the needs of the present without diminishing the options of the future generations still remains the authentic one. Initially concerned with conservation of resources it has now encompassed within its orbit the concern for environmental protection also.

Sustainable Development : Problems and Prospects

Many analysts blame environmental crisis on poverty-population linkages. They maintain that the teeming millions in India are responsible for environmental degradation. The poor masses allegedly degrade the environment by their uncivil and unscientific ways of using environmental resources to meet their needs for food, fuel and fodder. Even more importantly, they do so by multiplying their numbers and bringing the already scarce environmental resources under further strain. As such, poverty abetted by ever growing population is the worst pollutant. The prescription that follows this diagnosis is population control and slowing the pace of growth.

Without denying the linkages between poverty and population, it is essential to point out that this formulation does not capture the entire complexity of the economy-ecology inter-connections. In the first place, it presents a prejudiced view, in that it ignores the pernicious effects of prosperity on environmental deterioration. If one cares to audit the per capita profligate consumption of the natural resources and sinks in other developed areas of the country, prosperity emerges as the biggest cause of ecological decay not only in the developed but also in not so developed or underdeveloped areas of the country. Secondly, it glosses over the role of commercial sector that alone is responsible for the actual felling of trees for profit making in India as also probably in other countries. Thirdly, it involves the fallacious assumption that poor people tend to produce children in order to use them as workers. This is far from true. It is not because children can be used as workers that they are produced. It is because they are produced that they are used as workers. Fourthly, It is premised upon a truncated view of poverty that is mistaken as autonomous of equity. Indeed, poverty cannot be treated as an independent entity, separated from equity. Finally its prescription of slowing the pace of development is misleading because environmental degradation is the result of not excessive but insufficient economic growth.

Little wonder, there has come up an alternative explanatory formulation that seeks to account for environmental deterioration in terms of political economy. Recently articulated by some eminent scholars, this formulation attributes environmental degradation not to growth *per se* but to inadequate growth. It is argued that the Indian State has followed largely a capitalist path of development, which is evident from the policy choices that have been made in the spheres of irrigation, energy and forestry. Policy options in favour of large dams over small, in the sphere of irrigation, kerosene and diesel, over electrification of rural households in the sphere of energy, restrictions on collection of fuel wood and fodder for domestic consumption by the tribal and rural poor over planting trees on a massive scale in the sphere of forestry, are some of the instances in point. These policy decisions have contributed heavily to environmental contamination. The same can be said about the public policy that tends to favour private transportation over public transport, even as the former is far more polluting than the latter.

These policy choices are clearly indicative of an uneven pattern of development. Accordingly, our development has degenerated into a process of the state sponsored subsidized flows of resources to industrial elites and influential landlords. They have received water, power, forests and mineral resources, all at highly subsidized rates along with license to pollute the commons. As it is, three elite sections of our society, i.e. industry, bureaucracy and the politicians seemed to have collaborated to liquidate the country's base of natural resources.

The future too does not seem to hold any promise. In fact, it is believed that the policy of Structural Adjustment Programme (SAP) together with emphasis on privatization and globalization is likely to further aggravate the environmental problems, for there are critical areas of conflicts between liberalized trade and environmental standards.

Largely convincing, this explanation seems to over-estimate the role of policy choices in environmental decay. With the adoption of structural adjustment programme in 1992, there is no doubt a reversal in India's economic policy from regulated market to free market has led to aggravation of environmental problem. However, it remains to be noted that our environment was in a state of continuous degradation in the decades of 1970s and 1980s. Also, it was a period of regulated economic growth in favour of the poor masses. Similarly, the Indian State has taken definite policy choices in favour of environmental protection and reiterated its pronouncement of the same following the Earth Summit, but the state of India's environment continues to witness progressive decline. Therefore, even as policy choices are important, they are not adequate to explain environmental decline in India. There seems to be something more to it than mere policy choices. District administration in India is completely devoid of the idea or concept of sustainable development, while it still continues to inaugurate new projects and launch new programmes. Largely it is seen that whenever policies are planned at the state level, policy planners do not consider at all geographical or environmental conditions of that area.

According to Mahbub-ul-Haq, the author of the Human Development Report for the last few years, now the district administration and the Governments at the state level lay stress on the following imperatives: primarily top managerial approach over bureaucratized approach, promotional measures over regulatory mechanisms, natural resources based planning over finance based planning, and investment in human capital over physical capital, which so far has remained only a book maxim still to be realized.[6]

According to Haq's perspective, the first pre-requisite of sustainable development is de-bureaucratization and adoption of managerial approach. One reason for the disappointing performance of our family planning programme is its bureaucratized character. Likewise, it is bureaucratic approach to our programme of poverty alleviation that accounts largely for their uninspiring performance. Chief Medical Officers and other medical and health officers of the states have been entrusted with the task of realizing family planning targets, which badly failed, because of insensitive and impractical approach to the issue of family planning. The stories of official collusion with structures of vested interests, leading to a large-scale deforestation and blatant violation of antipollution acts, are all too well known. If the experience of East Asian countries is any guide, management approach holds promise for population reduction, economic growth as well as environmental regeneration. The same holds for primacy to promotional measures over regulatory mechanisms. The countries of South Asia, particularly India, excel at enacting legislation especially prohibitive legislation, only to observe it in breach. What is the use of having elaborate legislation if it cannot be enforced? Positive incentives are any day better than sanctions. Hence the importance of promotional measures for sustainable development must be understood.

Defining sustainable development as efficient optimization of natural resources, the 'efficiency' perspective further favours natural resource-based planning over merely financial resource-based planning. There is very little of natural resource accounting, much less environmental accounting in the country, which is why we continue to stay poor and deplete our natural resources unmindful of its consequences.

Investment in human capital is regarded as another critical input for sustainable development under the 'efficiency' perspective. Investment in health and education, particularly women's health and education, are known for their potential for sustainable development. There has come up cumulative, nay conclusive, evidence of correlation between female education and fertility decline. Spread of education, especially female education together with environmental literacy is thus likely to have multiplier effect on population reduction, hygiene, sanitation and environmental preservation.

Like the 'efficiency' perspective, the 'equity' perspective also, plays its important role in sustainable development. The economic successes in few of the states in India were made possible because of land reforms, the relatively low gap between high and low income and asset groups, high levels of education and literacy, better health and high levels of domestic savings and capital accumulation. Clearly, the stress here is on equity as an imperative of sustainable development.

Valuable in their own ways, both the perspectives have limitations of their own. The efficiency perspective for example, suffers from limitations of background, scale and priority.

Development model has succeeded in those states that gave precedence to development over free democracy in the formative phase of their economies. Even as they have had democracy, more often than not, it has been accompanied by authoritarian discipline.

The 'equity' perspective, on the other hand, is too ambitious to be workable. It calls for radical restructuring of class and power relations. It demands a different kind of political economy, a radical socialist economy, which has somehow failed to gain ground and has recently fallen in disrepute, over the matter of caste and communal clashes, centered around local politics. The district administrations have tried land reforms, but failed to enforce them, particularly the Land Ceiling Act. It also strived to remove poverty but ended up with widening the disparities between rich and poor. The state government and local administration aimed at creating a socialistic pattern of society as per the lofty ideals of our Constitution, but found traversing the capitalist path of development after 1991. Given the kind of soft state that is ever willing to submit to the designs of entrenched interests and/or to populist pressure, the 'equity' perspective has little chance to succeed.

Towards The Indigenous Perspective

The trouble with these perspectives is that they draw upon the lessons from East Asian and/or Western countries for devising a strategy of sustainable development for ourselves. Little do they care to look for a strategy within the native heritage and in the light of region-specific problems.

Some of the major problems of development common to all states of our country are poverty, unemployment, inequity, population growth, malnutrition, illiteracy, rural exodus and urban slums. Our development endeavor has added to this list a number of other problems, including environmental pollution, ecological degradation, marginalization of indigenous people, displacement of rural and tribal folk from their natural habitat and disfiguration of local cultures. Clearly, more problems have been generated than solved by the model of development that we have followed, which is indicative of there being something fundamentally wrong with it.

The flaw in the development model adopted by us lies in its wrong set of priorities. It has not paid sufficient attention to the optimum utilization of natural and human resources our region is endowed with. Rich in land and other resources, India has all along been primarily an agrarian society. If only we could harness our natural and human resources in a systematic and rational manner through proper manpower planning, we would not have faced the problems of exodus from rural to urban, over-congestion and slums, marginalization of the indigenous people and so on. No doubt we have made efforts to modernize and diversify agriculture, but have not done enough to maximize its full possibilities. As a matter of fact, eastern Uttar Pradesh in general and Gorakhpur in particular has a tremendous potential to become a leading supplier of agricultural and agro-based industrial products apart from feeding its own population. This strategy has a cultural content as well in that it is in complete accord with the socio-cultural gestalt of the region.

Living in harmony with nature has been an important feature of the cultural heritage of India. It is a pity that this healthy tradition has been ruptured by the faulty model of development that was adopted. Rather than destroying environment, including biodiversity, we should learn to live in harmony with nature, as postulated in our indigenous vision.

Another dimension of our culture, which has crucial relevance to sustainable development is the primacy of socio-cultural values over materialistic interests. In the prevailing paradigm of development, this relationship has been reversed with material interests taking precedence over socio-cultural values. This distortion precisely is at the root of many of our development and ecological problems. Sustainable development cannot be achieved by means of policy choices alone, no matter how enlightened and well thought out these are. The restoration of certain basic human values such as civility, discipline, hard work, professional integrity and, above all, concern for others, is necessary for environmental restoration and ecological preservation. The current strategy of value-based development in order to pursue the goal of sustainable development is of paramount importance.

It is high time that we bring to bear the integrated perspective on the notion of sustainable development. In the West, sustainable development has been conceptualized in a segmental manner. This is evident from the use of phrase such as ecologically sustainable, politically sustainable, socially sustainable and culturally sustainable development. This compartmentalized way of defining sustainability is fallacious, as it implies autonomy of one sphere vis-a-vis the other, which is not the case in the living reality. Logically speaking, the idea of segmentally sustainable development is contradictory in terms. For example, it sounds paradoxical to speak of sustainable development, if it is sustainable ecologically but not culturally.

IS SUSTAINABLE DEVELOPMENT DESIRABLE AND POSSIBLE

Is sustainable development desirable? Does it fulfill the goals currently put topmost on the human agenda: equality, happiness, justice, peace? Or is it a contradiction in terms, with development being inherently unsustainable?

Why Are We Where We Are?

An answer to these questions will depend on what is thought to be 'sustainable' and what is meant by 'development'. I cannot but express my unease by the way this phrase is branded about by everyone from the neighbourhood development consultant to the Indian Prime Minister, not to mention the President of World Bank. Admittedly, the concept of sustainable development originated in the fierce debates raised by environmentalists in the last two-three decades, but its recent co-option by the very forces which are destroying the environment, is an occasion for us to take a step back and re-examine its meaning and validity.

It is obvious to anyone with even a bit of sensitivity that humanity is dreadfully out of tune with its surroundings. In the last couple of centuries, it has blundered about like a drunken bull in china shop, ripping to bits the fragile natural web on which all life depends. I will not repeat here the dreary and by-now cliched facts regarding the extent of damage caused to the earth by human hands, and the ways in which this damage has rebound on us. A society which condemns its new-born members to high doses of pesticides in their mothers' milk, cannot but be called perverted. Not to mention the fact that the same pesticides also end up damaging the health of penguins thousands of kilometers away, carried there by complex ecological inter-connections, which are beyond the grasp of our development planners.

The notion of development therefore, needs closer examination. It has been forgotten that the term is the opposite of 'envelopment' and signifies an 'opening out', and expansion of possibilities. Therefore, development would include not merely material growth, but also, and perhaps more importantly, the blossoming of cultural, intellectual, spiritual and other non-material aspects of human existence. Unfortunately, the Western notion of development has sidelined all these aspects, and brought materialism to the centrestage. Even intellectual and cultural growth today seems to be geared towards the aim of making money and accumulating goods. The contradictions and limitations of such an approach are very well typified by that seminal work, *Our Common Future*, which is increasingly being adapted by countries in their National Planning.

OBJECTIVES OF THE RESEARCH

The present work is aimed at realizing the following objectives:

1. To assess the development programme of the area, that is one of the most important district headquarters of Eastern Uttar Pradesh.
2. To understand the policies, principles and programmes adopted by district administration for promotion of sustainable development.
3. To study the habit patterns, social and cultural activities of the local population in order to understand their approach towards sustainable development.
4. To evaluate economic and other livelihood related professions of the local population so as to understand their sustainability in terms of environmental and ecological balance of the area.
5. To make an assessment of the level of awareness among the masses as well as officials about the concept and approach of sustainable development.

The underlying theme and objective of the present research is to study and examine the delicate and fragile relationship between environment and development of an area of this part of eastern Uttar Pradesh which has been politically so fertile but economically so depleted.

RESEARCH METHODOLOGY

In the present work standardized methods and techniques of public administration have been followed. It consists of observation and interview for the primary data collection. For secondary data collection, libraries, official documents, district gazetteer, review of literature etc. has been followed.

Observation has been quite fruitful in data collection in the sense that it is a first hand data collection technique, where intermediary biases are eliminated. However, it has been very difficult many a time because officials accompanied me during my observations. They usually provided explanations continuously and tried to fabricate the facts. Further, observation was helpful in cross checking the data taken through the interviews.

Interviews are the standardized technique followed in the discipline of public administration. I have made extensive use of this technique. Interview with the officials such as Commissioner of Gorakhpur region, District Magistrate, Chief Development Officer, Pollution Control Board Officials, District Panchayti Raj Officer, Block and Village level workers along with City Commissioner of Nagar Nigam and Vice Chairman, Development Authority provided me fruitful and vital information regarding government programmes on sustainable development. It also helped me to evaluate the implementation of the programmes. Interview with the local people, community leaders, members of local self government institutions, workers, other professionals, farmers and daily wagers immensely helped me in getting their responses towards the government programmes. However, many problems cropped up during interviews. The government officials usually gave a lukewarm response and remained indifferent while answering to my questions. Most of them hardly remembered the details of the programmes. I was also surprised to see that many of them had no idea about the concept of sustainable development. Their answers were vague and often misleading.

On the other hand, among the local people some reacted very sharply against the apathy of the government departments and had an extreme sort of approach even without knowing what sustainable development stands for. Some were neutral, still some other remained dull with no response at all. Many local leaders tried to raise their own concerns and aspirations even without acknowledging the question I had asked. I had to set the situation with good rapport to elicit the information from the local people.

Review of literature provided the basic framework for the present work. It gave me ample opportunity to develop a perspective and structure of research. I gained concepts and ideas while going through books and articles of eminent scholars. Use of libraries was especially helpful in going through various periodicals and journals at Delhi School of Economics, Jawahar Lal Nehru University, in New Delhi and the Secretariat Library of UP state assembly and Tagore Library along with Department of Public Administration Library of University of Lucknow. The extensive search of materials, facts and figures gave me sufficient inputs to build up this work.

I also analyzed official documents from various government departments, especially, development departments, Municipal Corporation, development authority and the district gazetteer. The review of literature and the observations through interview, both were verified, thus enabling me to develop basic structure of my present work.

After collecting data and information, these were systematically analyzed in a standardized manner through tabulation and interpolations. Since the data were mainly qualitative in nature, the analysis also remained qualitative. In this manner, the whole research gave me ample insight of sustainable development. In the concluding observation I have also put forward certain suggestions based upon gathering of my knowledge on sustainable development of the Gorakhpur district, the area of my case study.

NOTE

1. United Nations Development Programme, *Human Development Report, 1991*, New York, Oxford University Press, 1991.
2. L.Goodland and G. Ledec, "Neo-classical Economics and Principle of Sustainable Development", *Ecological Modelling*, Vol. 38, 1987.

3. World Conservation Union, *Caring for the Earth*, Gland (Switzerland), UNEP, 1991.

4. Quoted by Bamberger and Cheema, *op.cit*.

5. Hans-Jurgen Harborth, "The Debate about Sustainable Development: Starting Point for an Environment-Oriented International Development Policy" *Economics*, (ed. by the Institute for Scientific Cooperation, Tubingen) Vol. 44, 1991, pp.10-15.

6. Mahbub-Ul-Haq and Paul Streeten, *Human Development Report 1990 UNDP*, New York, Oxford University Press, 1990.

2
The Setting

As mentioned earlier, the study has been carried out at the district headquarters of Gorakhpur, a very prominent area in political, economic and geographical terms in eastern Uttar Pradesh. The area has been chosen for the research because of its uniqueness for Uttar Pradesh in general, and eastern Uttar Pradesh in particular. At the same time it has been my home district before Maharajganj was carved out of it. Gorakhpur is politically very fertile but economically the most barren district. It is highly unfortunate that not a single development programme in the district is based upon the basic elements of sustainability. Neither the parameter of ecology and environment has been taken care of nor the officials are aware of it.

A general outline about the district is given as such.

Profile of Gorakhpur District

Gorakhpur District is situated between 25″ 51′ north and 26"30′ south Latitude and at 83″25′ east and 84″24′ east Longitude. The name 'Gorakhpur' is based upon a famous Saint of 'Shaive Cult' Shri Gorakshnath.

In 1806, the area of Gorakhpur was transferred to the East India Company from Nawab of Avadh. At that time its northern border was extended up to Butwal in Nepal. Its eastern border touched the state of Bihar and in south, it was surrounded by districts Jaunpur, Gazipur and Faizabad. In the west,

Gorakhpur was surrounded by districts Gonda and Bahriech. At that point of time Gorakhpur was composed of 17 parganas of the districts Basti, Azamgarh and Mahool, containing 6 parganas of Nawabgang and Gonda, 7 pargans of Khairigarh, which was situated in the north of district Khiri and also the area of terrain of Binayakpur and Tilpur were the parts of Gorakhpur district. In 1802, 7 parganas of Khairigrah were separated. In 1816 after the famous battle of Butwal, by an agreement, Binayakpur and Tilpur were handed over to Nepal. In the same year parganas of Nawabganj were again included in the province of Avadh. But in 1820, most of the parts of Azamgarh and Mahool were transferred to Jaunpur and Gazipur. In 1865 most of the parts of Maghar pargana and Binayakpur pargana were merged together to give rise to a new district named Basti. In 1904 Ghaghra River changed its course, as a result, 122 villages of Dhuriapar pargana had to be transferred to Azamgarh district because of administrative point of view. In 1946, district Gorakhpur was again divided to give rise to a new district named Deoria and again in 1987, a part of Gorakhpur and Deoria districts were merged together to form a new district Maharajganj. At present, the northern border of Gorakhpur district touches Maharajganj and in the east with Deoria and Kusinagar. In south, river Ghaghra acts as a border across which districts Mau, Azamgarh and Ambadekar Nagar are situated. In the west of Gorakhpur is situated the districts of Sant Kabir Nagar and Siddharth Nagar.

At present, in Gorakhpur district there are 7 Tehsils namely Sadar, Kampierganj, Chauri Chaura, Sahajanwa, Khajni, Basgaon and Gola. The main rivers that flow through the district are Rapti, Ghaghra, Rohin, Aami, Kuwano and Gorra. These rivers often swell after the monsoon, badly affecting the district through floods, resulting in huge losses of men and material.

The inclination of land in the district is usually towards the south from north. During rainy season rivers and *nalas* of the district assume turbulent shape, which badly affect the agricultural property and development activities. Because of frequency of floods, quality of soil also keeps on changing. As a result, agriculture production is affected.

District Gorakhpur is one of the most significant religious, cultural, educational and trade center of eastern Uttar Pradesh. Gorakhpur is situated at the confluence of river Rapti and Rohin. The average height of the district from the sea level is 185 meters. Its internal structure and composition has been affected on number of occasions with the change in the course of river Rapti from time to time. Geographical area of the district and the city has also been influenced many times by the ravaging floods. The city Gorakhpur is bowel shaped. As a result, there is a lack of natural water drainage.

DISTRICT PROFILE OF GORAKHPUR

Gorakhpur is situated at Lucknow-Barauni main Railway track and is now the head quarter of North Eastern Railway, therefore, it is one of the most important railway junctions of the eastern Uttar Pradesh. It is almost 90 kilometers away from 'Indo-Nepal' border. The district headquarters of Gorakhpur houses a commissionerate. As far as population is concerned, it is only next to Varanasi in the whole of eastern Uttar Pradesh. (See table 2.1)

In 1954, the North Eastern Railway headquarter was established here and in 1964 a bridge was constructed on 'Rapti' river. Thus, Gorakhpur was connected with National Highway number 28 and other state highways. In this way, Gorakhpur is now very well connected with road and railway links. As a result, the city has come up as one of the important trade and commercial centers of eastern Uttar Pradesh. The district has a Railway workshop, the famous 'Geeta Press', now a closed fertilizer factory, one engineering college and the medical college, one university, many post graduate and degree colleges, a security unit of the Indian Air Force and centres of 'Akashvani' and 'Doordarshan'. Gorakhpur Industrial Development Authority (GIDA) has been set up to speed up the process of growth and industrial development.

In 1991, the decadal growth of population in the city was more than 3.21 percent and is, therefore, now one of the corporation-towns of Uttar Pradesh. It has been proved to be a source of inspiration for the people of surrounding districts,

Table 2.1 :

1.	Geographical area	3483.8 sq.km.
2.	Population (Based on the census of 2001)	3784720
	a. Male	1931782
	b. Female	1852938
	c. Rural	2451917
	d. Urban	1332803
3.	Ranking by Population size	5th District in Uttar Pradesh
4.	Decadal Growth rate	23.44%
5.	Sex Ratio	959
6.	Ranking by sex ratio	11th in Uttar Pradesh
7.	Population Density	1140 per sq.km.
8.	Ranking by Population density	7th District in Uttar Pradesh
9.	Literacy rate	60.96%
	a. Male	78.70%
	b. Female	44.45%
10.	Tehsils	7
		(Sadar, Chauri Chaura, Sahjanwa, Khajani, Kampierganj, Basgaon, Gola)
11.	Blocks	19
		(Jungle Kaudia, Kampierganj, Khorabar, Chargava, Bhathat, Pipriech, Pali, Sahajnwa, Sardar Nagar, Rampur, Piprauli, Khajni, Belghat, Urva, Barhalganj, Gola,Basgaon, Kauriram, Gagha,Nyayapanchayt of Dhani Block and one Panchayt of Farenda).

Contd....

Table 2.1 : (Contd...)

12.	Police Stations	25 (Kotwali, Rajghat, Tewaripur, Gorakhnat, Shahpur, Gulriha, Chilutal, Khorabar, Basgaon, Urva-bazar, Gola, Belghat, Pipriech, Cahuri Chaura, Jungha, Khajni, Sikiriganj, Sahajanwa, Harpur, Budhat, Badhalganj, Gagha, Belipar, Kampierganj, P.P ganj).
13.	Nagar Nigam	1 (Gorakhpur)
14.	Nagar Panchayat	7 (Badhalganj, Gola, Basgaon,P.P ganj, Piprieach, Munderva Bazar).
15.	Nyaya Panchayat	191
16.	Gram Sabha	1233
17.	Villages	3319
	(Inhabited)	2876
	(Uninhabited)	439
	(Forest village)	4
18.	AgricultureMandi committee	4 (Gorakhpur, Sahajanwa, Chauri Chaura, Mundera Bazar.)
19.	University	1 (Pundit Deen Dayal Upadhyay Gorakhpur University.)
20.	Engineering College	1 (Madan Mohan Malviya Engeenering College)

Contd....

Table 2.1 : (Contd...)

21.	Medical college	1 (Baba Raghav Das Medical College)
22.	Degree Colleges	15
23.	Higher Secondary Schools	135
24.	Senior Basic Schools	363
25.	Junior Basic Schools	1413
26.	Aganwari and Balwari centers	510
27.	Allopathic Hospitals	22
28.	Ayurvedic Hospitals	44
29.	Homeopathic Hospitals	24
30.	Unani Hospitals	2
31.	Primary Health Centers	57
32.	Veterinary Hospitals	45
33.	Railway Stations with Halts	20
34.	Bus stations/Bus stops	106
35.	Ration Shops	2104
	(In Rural areas)	1797
	(In Urban areas)	307
36.	Nationalized Banks	113
37.	Gramin Banks	60

Contd....

Table 2.1 : (Contd...)

38.	Co-operative Banks	32
39.	Co-operative Agriculture and Rural Development Banks	4
40.	Total Cultivated Area (in Hectares)	26600
41.	Area Cultivated More than Once (in Hectares)	12400
42.	Net Irrigated Area(in Hectares)	18900
43.	Length of Canals(in Kilometers)	42.2
44.	Government Tubewells	894
45.	Private Tubewells andPumpsets	102191
46.	Length of the	
	Railway Track	109 km
	Broad Gauge	53 km
	Meter Gauge	56 km
47.	Rural Post Offices	324
48.	Urban Post Offices	663
49.	Total length of Metalled Road	1730 km.

Source : District Gazzetter.

such as Deoria, Padrauna, Maharajgang, Basti, Siddharth Nagar, Azamgarh and Mau.

The climate of the district is affected by the monsoon and in the month of May the average belt temperature of the district is 38.4 degree centigrade maximum and in the month of January 7.2 degree centigrade minimum. The average monthly rainfall of the district is 1270 mm. Generally direction of the wind is from east to west. The air contains humidity almost through out the year and because of lack of *'Loo'* the heat is bearable in summer.

Geographically Gorakhpur is situated in that fertile land of 'Ganga-Yamuna' where the most poverty striken population lives in. The Gorakhpur city has a fewer number of big industrialists, commercial masters and landlords, rather most of the population is poor, land less and illiterate labourers. A big section of the population earns livelihood as daily wage earners or labourers not only in and around the district but also in far off places like Mumbai, Surat, Kolkata, Chandigarh, Ludhiana, Delhi and other far flung areas. Most of the people are temporarily employed in various other provinces in the country. In the last two decades a major chunk of native population had even migrated to foreign locations such as Bangkok, Trinidad and Tobago, Mauritius etc. Nature has endowed the district with the conditions of agricultural facilities. Not only this, but also the favourable climate, fertile soil, sufficient rains, perennial rivers and higher level of groundwater make the district most conducive for agricultural activities. But, the secluded location and non-connectivity with the railway link for a very longer period of time in the past, except for the narrow gauge railway link, led to the lack of industrialization in the district. A few sugar mills are exceptions that are now running in losses. In the decade of 1980s, district Gorakhpur was notoriously famous for criminal activities of dominant classes namely Brahmins and Thakurs. Consequently, industrial development and growth was badly affected specially from 2nd to 7th five-year plans. When in all other areas of the state industries were established, Gorakhpur had to contend only with one fetilizer factory.

Like all other ancient cities of the country, Gorakhpur is also clearly divided into two parts; 'Old Gorakhpur' and 'New Gorakhpur'. The Old Gorakhpur has unplanned internal structure like other ancient urban centers having narrow lanes and by-lanes, dense population and ancient buildings with improper drainage and sewage disposal system. Old Gorakhpur is located in the centre of the city with lack of infrastructure. Later on, residential colonies were developed which now make major parts of the main city. New Gorakhpur is centered around its main locality Golghar and Civil-line areas which developed after independence, is rather more planned. This part contains less density of population in comparison to Old Gorakhpur. The new residential colonies, such as Suraj Kund, Betiya Hata, Rajendra Nagar, Priyadarshini Nagar, Sahara Housing Estate and Railway Colony are better planned and are developed in grid construction. Modern development of the city is now taking place on sides of the main roads of state highways leading to different district headquarters.

Like all other developed cities of the country, Gorakhpur does not have different functional sections except a few administrative offices. The land use of some areas in the district has led to the industrial urbanization of the city. In Gorakhpur city, modern markets centres like Urdu Bazar, Shaibganj, Alinagar and Ghantaghar have made it one of the important trade centres in Eastern Uttar Pradesh. Major trading and commercial activities take place in-and around road shops of Golghar, Basharatpur, Vijay Chowk, Ghosh Company, Halsiganj, Pandey Hata and from Alinagar to Raiganj, from Shahpur to Lal Diggi, Khoonipur and Chauriagola. Apart from this, Mohaddipur, Kuraghat, Ashuran and Gorakhnath Road are other major trade centres of the city. To cater to the needs of the local population, there are various shops and small markets in almost every locality.

Infrastructure of the district is pathetically poor. As a result industrial and other commercial activities are absent in the town itself. There are thirteen major industries while the rest are medium or small-scale industries. All these are located in densely residential areas except the Fertilizer Corporation that is 8 kms. away from the district collectorate on Gorakhpur-Nawatanwa Road. On the pattern of NOIDA, to speed up the process of industrialization in the whole of eastern Uttar Pradesh, Gorakhpur Industrial Development Authority (GIDA) was established in 1989, which unfortunately fell prey to peasants' movements, lack of facilities, financial discrepancies and administrative mal-assistance. As a result, development in Gorakhpur is most unfriendly and unplanned. As far as ecology and environment is concerned, elements of sustainable development are negligible and Government officials are either insensitive or unaware of the concept of sustainable development.

Open lands, play grounds, parks and proper sewage disposal system, which in-fact are considered to be significant indices of civic amenities and sustainable development are either badly in a pathetic state or absent in the area. At present there are only six parks: Lal Diggi, Town Hall, Premchand Park, Betiya Hata, Rajkiya Udyan Park and Mohaddipur. Apart from this, there is one regional stadium and a Parade Ground near Railway Station in the east. There is a Golf ground spread over 22 acres of land in another part of the city. There is a lack of open place in the city.

In 1989, in the mid-south of the city, one Ram Garh Tal Project was initiated to develop Gorakhpur as one of the major tourist centres. Spread over an area of more than 169 acres, the project could have proved to be a boon for whole of the Gorakhpur district but due to lack of financial resources, intermittent political interference and apathy of the Government, fate of the project is uncertain.

Gradually, in due course of time, population from and around neighbouring districts and villages started pouring into the city area, giving rise to plethora of problems related to urbanization and expansion. As a result, on 15 January 1977, Gorakhpur Development Authority was established under the Uttar Pradesh Urban Development and Planning Act, 1973 that actually started functioning in 1986. So far, there are 11 housing schemes of the authority such as Charu Chandra Puri, Rapti

Nagar, Jan Priya Vihar, Surya Vihar, Vikas Nagar, Vikas Nagar Extension, Shastri Nagar, Shastri Puram, Railway Station Residential Scheme etc,. Perusal of these schemes make it amply clear that the development authority is mainly engaged in constrution activities that too without any care to environment and ecology of the area.

As a long-term plan of sustainable development and growth, 162 projects worth Rs. 970.39 lakh in the financial year of 2001-2002 has been proposed for the district. 11 are related to sewage and sewage disposal, 22 with water supply, 123 with road construction, 6 with computerization and 2 schemes with that of electrification. Whereas, in the financial year 2000-2001, a total of 244 projects were sanctioned to Gorakhpur Municipal Corporation worth Rs. 651.67 lakh, out of which 8 projects are related to water supply, 5 to roads, 193 to solid waste management, 3 to parks and beautification, 4 to anti-encroachment, 9 to computerization, 3 to miscellaneous and 2 with that of electrification.

TOPOGRAPHY

The district presents characteristics distinct from natural features of the western districts, of the State. This difference is primarily due to the relative proximity of the Himalayas. The outermost foothills are but a few kilometres distant from the northern borders, and the high peaks of the snowy range, culminating in the huge mass of Dhaulagiri, some 8,230 metres above sea-level, are clearly in sight under favourable climatic conditions as far south as Gorakhpur itself. Below the outer hills is a dry boulder-strewn tract, corresponding to the Bhabar of Kumaun and Garhwal and here the bulk of the moisture contributed by the rainfall and the small streams is absorbed by the soil, to reappear through seepage in the damp and unhealthy tract, known as the terai. The latter comprises a belt, some 16 km. in width, running along the northern borders of the Maharajganj district. It is extensively cultivated. In south of the terai in the same district is a stretch of forest land, which extends downwards in patches as far as the centre of the district. The average depth of water is about 4.5 meters. The plains form a level tract, which slopes gently from west to southeast. The height above sea-level ranges from 107 meters in north-west to 93 metres in the south east. Higher elevations appear at places where the general flat surface is broken by irregular ranges of sand hills. The most clearly defined ridge of this nature starts near Hapur in the Maharajganj district and runs in a winding course almost to Deoria. It presumably marks the long abandoned channel of the Gandak or some other river, since, throughout its length, it is bordered by a chain of depressions and *jhils* and in several places pebbles and boulders have been encountered in sinking shafts for wells. In contradistinction to the high ridge are the low and often broad valleys of rivers known as *kachhar*. The valleys of the larger rivers are not only depressed well below the general level of the country but are of considerable breadth. Thus there is a wide area of low land, which is inundated in years of heavy rainfall.

River System and Water Resources

The main river system known as the Rapti system is confined to the west side of the Gorakhpur city. The valleys of the Ghaghra, the Rapti, the Rohini, and the Ami at any rate in their lower reaches, are at places broad and sufficiently depressed below the ordinary level of the district and confine their floods within the limits of the high banks on either side.

The drainage of the entire district, excepting that carried off by the Great Gandak, is discharged into Ghaghra. In many places the drainage is imperfect especially in the basin of the Rapti and its affluents.

Rapti—The Rapti, originally Irawati and then corrupted as Ravati, has its source in the outer ranges of Nepal. After traversing Bahraich, Gonda and Basti it enters this district between Talnatwa and Bersar and after forming the boundary for a few kilometers, flow in a south- easterly direction. It again forms the boundary from Rigauli to Jainarayanpur village between Basti and Gorakhpur districts. Thereafter it moves southwards and traversing about 20 kms. It turns towards east and touches Gorakhpur city, and again takes a turn towards south-east. It then flows in an extremely tortuous course and flowing southeast it forms the boundary between Basgaon tahsil and district Deoria leaving the district in the extreme southeast corner of tahsil Basgaon. The river is constantly carving out new channels for itself and occasionally shifts whole villages from one bank to another. Occasional floods leave behind them extensive deposits of silt and sand which are generally beneficial, since the silt is of an exceedingly fertile nature, while even the sand after two or three years' exposure weathers into a loam capable of bearing good crops.

Ghonghi—The tributaries of the Rapti are numerous and important. The first to join the river on its left bank is the Ghonghi, which originates in the outer ranges of hills above the Napalese terai and for some distance the midstream of Ghonghi forms the international boundary between India and Nepal. Flowing in a southwesterly direction, it receives on its left bank, two small terai streams called Dands and Dunri. Then it turns south and for some distance provides Basti boundary, flowing past Bridgmanganj. Subsequently it splits into two channels, both of which unite with Dhamela at Sikri and Gerui near Riguali respectively. It has a deep and well-defined channel with a sandy bed and clear water. Though usually much swollen during the rains, it shrinks rapidly and soon becomes fordable. The Dhamela is merely an old channel of the Rapti, which receives the water of the Kunhra and other rivers of Basti terai, and after a course of some 15 kms. through this district rejoins the original bed of the Rapti just above Karmaini.

Rohin—The next affluent of the Rapti is the Rohin, a stream of some magnitude which enters pargana Binayakpur form Nepal and after traversing pargana Haveli in the southerly direction, falls into the Rapti between Domingarh and the western extremity of Gorakhpur city. At first its banks are steep and the bed is formed of sand and pebbles, but on its entry into pargana Haveli it loses these characteristics and approximates more closely to the ordinary type of river in the plains. The Rohin receives about 9 kms. southwest of Sakhuani on its left bank, the Baghela, which is in turn fed by the Madrahi and other streams. The Piyas or Jharain enters this district at Thuthibari and is joined by the Malaun river rising near

Lohrauli and the Nadao Nala and other small watercourse before falling into the Rohin just above the crossing on the road between Pharenda and Maharajganj. The river Balia rises near Maharajanj and falls into Rohin near Jarlahia. The river Chillua takes its origin in *tappas* Katahra of pargana Haveli and is fed by the Temar and some minor streams before passing into the Chillua Tal and then into the Rohin near Maniram. On the right bank, the only affluent is a water course called the Kalan which rises near Lehra and flowing southwards joins the Rohin in the north-east of P.P ganj.

Tura and Gaura—The Tura is a small stream which rises in *tappa* Unit of pargana Haveli and flows southward through the Ramgarh forest to the east of Gorakhpur city till at the village of Jhangha it joins the Gaura. The latter carries off the overflow from the Ramgarh and Narhai Tals, and the combined stream continues for a considerable distance parallel to the Rapti.

Pharenda Nala—Rising in the north of Pipraich flows southwards almost parallel to river Tura and finally joins Gaura in the south of Rajdhani. The name Pharenda is derived from the thick growth of 'Pharenda' of wild *jamun* trees along its course.

Ami—The tributaries of the Rapti on its right bank are Ami and Taraina. The former rises in pargana Rasulpur in Basti district and enters this district near Rampur in pargana Maghar (Sant Kabir Nagar District). Then it flows towards southeast falling into the Rapti near Sohgaura in pargana Bhauapar. It is a narrow and sluggish stream saved in rains when its swollen waters develop into the Amiar Tal. The spill from the Rapti during heavy floods enters the Ami valley, submerging the country as far west as the high right bank, save for two islands of *bangar* which always remain above the flood level. One stretches from Kalesar to Newas along the south side of the Basti road, and the other lies on either side of the Azamgarh road from Hardia to Malaon.

Taraina—This river takes its rise in the south of paragana Unaula and flowing in a south-easterly direction through Dhuriapur reaches to Bhenri Tal. Emerging from the eastern extremity of that lake it finds its way into the Rapti by an almost direct route due east. During the hot weather it is an insignificant stream but during rains it assumes considerable dimensions.

Ghaghra—The Ghaghra, also known as the Sarju and Dehwa, an accumulation of the combined water of the Chauka or Sarda and the Kauriyala, which unite near Bahramghat in the Bara Banki district. This is a mighty river flowing over a wide and sandy bed. The deep stream forms the boundary between this district on the north and Azamgarh on the south. This river touches Gorakhpur at Majhdip in pargana Dhuriapur and then flows towards east touching the market towns of Gola and Barhalaganj. The bed of the river is flanked by steep and well defined banks but occasionally the water rise in heavy floods, inundating the lower land in the neighbourhood. When the river subsides after the rains, numerous sandbanks and islands appear in the channel, which becomes tortuous and in places shallow.

Kuwana—Besides the Rapti, the other direct affluent of the Ghaghra in this district is the Kuwana. This river has its source in the east of Bahraich and it flows through Gonda and Basti. It separates Gorakhpur district from the Basti district, and then passes through the west of pargana Dhuriapur to join the Ghaghra near Shahpur. In this portion of its course it has a sandy bed with fairly steep high banks.

Great Gandak—This river takes its rise in the snowy range of Nepal and leaves the hill by the gorge near Tirbenighat, about 16 km. north of the boundary of the district Maharajganj. This river, which is also known as the Narayani, and in Nepal as Salilgrami, flowing in the north-east of district Maharajganj enters the Deoria district. The Great Gandak is a stream of the first magnitude and even in hot weather its volume is immense. On its first entry into the district the bed is stony and the water is cold and clear. The river is subject to violent and sudden floods, which cause extensive inundations of the forest tract in Nepal and in this district.

Choti Gandak Nadi— Rising in Nepal it flows southwards and enters this district at village Sitlapur. About a kilometer and a half south of this village it bifurcates into two branches, one branch running in a north-westerly direction and other continues on a southerly course as Choti Gandak and past the Domakhand forest tract reaches the Deoria district to its south. Flowing in the same direction it forms the boundary between district Maharajganj and tahsil Hata of district Deoria leaving this district in the extreme south-eastern corner of district Maharajganj.

Lakes— Gorakhpur has a number of large perennial lakes, formed in the most cases in the abandoned channels of rivers, which have become blocked by the accumulation of silt, or by the accumulation of water in deep natural depressions. The important ones are being mentioned here.

Ramgarh Tal— On the south-east of Gorakhpur city and on the south side of the metalled road to Kasia in Deoria district is the Ramgarh Tal. It was formerly covered, save during the rains, with a dense growth of reeds, which was a menace to the health of the town and has since a large part of it been cleared off. An unsuccessful attempt was also made to drain off the water from this lake through a channel from the southern end into the Rapti. Ramgarh Tal is rich in fish and affords living to people in several villages on its banks.

Narhai Tal— A few kilometer south-east of Ramgarh Tal a much smaller sheet of water lying in the *Kachhar* of the Rapti is known as Narhai Tal. It is connected with the Ramgarh Tal by the Gaura river which carries off the excess water of both the lakes into the Rapti. This lake was formerly a channel of the Rapti. During summer it becomes dry and serves as grazing ground for cattle.

Domingarh and Karmaini Tals— These lakes in the west of Gorakhpur city are formed by the overflow of the Rohin just before its confluence with the Rapti. The two lakes are separated by a mound which is completely submerged by water during high floods and the two, for all practical purposes become one. During rains the water extends northwards from the railway embankment at Domingarh over a continuous stretch of about 11 km. After the close of rains the water rapidly drains off into Rapti and the lakes shrink to small dimensions.

Nandaur Tal— About 9.6 km. south from Gorakhpur lying close to the east of the Azamgarh road, is the Nandaur Tal. It is a perennial sheet of water. Its size is little affected by the rains. The water is clear and it is an inexhaustible source of fish.

Amiar Tal—A few kilometers south of Nandaur Tal is Amiar Tal, formed by the flood waters of the Ami river which fill the whole of the valley between Belipar on the north and Bansgaon on the south, both situated on ridges. The intervening depression during rains becomes a lake several kilometres in length terminating at the tucker bandh. To the east of the bandh a second lake is formed known as the Bijra Tal. The water recedes quickly after the rains and the land thus exposed yields good *rabi* crops.

Bhenri Tal—This lake lies between the Rapti and the Ghaghra rivers in paragana Chillupar and is formed by the excess water of the Taraina river which passes through the lake. In the dry weather it shrinks to small dimensions but during rains it becomes a large expanse of water. This lake is however shrinking permanently owing to increasing accumulation of silt. A channel from the eastern extremity carries off the surplus water into the Rapti but, during floods, the lake overflow its banks, submerging the entire neighbouring area.

Chillua Tal— The Chillua Tal is formed by the overflow of the Chillua *nala* in pargana Haveli about 11 km. north of Gorakhpur. It is a long and somewhat narrow stretch of water shrinking into river near Maniram before joining the Rohin.

Other Lakes— In the eastern part of the district there are many lakes such as Ramabhar Tal, Kuseshar Tal, Chakahwa Tal and Dumrani Tal.

GEOLOGY

The geology of the district exposes nothing beyond ordinary river-borne alluvium, which is not old. The mineral products are few and unimportant. The minerals of commercial value are the nodular limestone conglomerate known as *Kankar*, brick earth and saltpetre. The last occurs principally in the south and south-east and is manufactured in a crude state in considerable quantities, most of it being exported to the markets in Bihar. In the Basgaon tahasil *kankar* is most abundant and quarries are seen at many places. It is also extracted from some places in Maharajganj district. Lime is obtained by burning *kankar*. Brick clay is abundant everywhere and bricks are made all over the district. The soil in the district is light sandy or dense clay of yellowish brown colour. The sand found in the rivers is medium to coarse grained, grayish white to brownish in colour and is suitable for construction purposes.

Seismology—In the earthquake zonal map of India the district lies in zone IV liable to moderate damage by earthquakes. Although no major earthquake occurred close to it, the tract being not far form the Great Himalayan Boundary Fault, experiences the effects of moderate to great earthquakes occurring there. The seismic intensity may not exceed VIII on the Modified Mercalli scale-1931.

FLORA

The forest are generally found in the northern portion of the district though in the past they extended as far as to the south of Gorakhpur and along the Rapti in southeastern part of the district.

About 55,235 hectare of land covered with timber and other trees and shrubs is under the management of the forest department. Of this an area of about 20,720 hectares lies in tehsil Pharenda, about 30,175 hectares in district Maharajganj and about 4,340 hectares in district Gorakhpur. The forest area in the district under the control of the Gaon Sabhas is about 1,490 hectares of which 626 hectares are covered with timber trees and the remaining with other species of trees and shrubs. Of the timber forests tehsil Pharenda contains 206 hectares, district Maharajganj 409 hectares and Gorakhpur 11 hectares. The forest area having other trees and shrubs extend over 30 hectares in tehsil Pharenda, 549 hectares in district Maharajganj, 166 hectares in tehsil Bansgaon and 119 hectares in district Gorakhpur.

The sal is the principal tree of the forests found in the district. The undergrowth in sal areas is very dense in the strips along the banks of *nalas* and streams, and is of evergreen shrubs and small trees. In the northern portion of district Maharajganj there is Domakhand forest tract. This has dense forests of medium to good height. The chief associates of the sal are *asna (Terminalia tomentosa), bahera (Terminalia bellerica), haldu (Adina cordifolia), phaldu (Mitrgayana pargifolila), fasidh (Ligevstroemia speciosa), tendu (Diospyros tomentosa), majua (Madhuca indica), domsal (Miliusa veluatina), Jamun (Syzygium cumini), bhfakmal (Ardisia solancea), Shisham (Dalbergia slssoo), semal (Salmlia malabarica), tun (Cedrela toona), dhak (Burtea monosperma), neem (Azadirachia indica), amaltash (Cassia fistula), imli (Tamarindus indica), and teak (Tectona grandis).*

The afforestation programme is being carried out in about 550 hectares annually. Sal and teak were planted in about 300 hectares in 1973 and *shisham, khair, semal,* etc., in about 200 hectares. In addition, the forest department has carried out plantation of roadside avenues on national highway from Gorakhpur to Lucknow for about 23 km.

Groves—The total area under groves in the district in 1973 was 13,729 hectares. The largest grove area lies in district Gorakhpur where it is 5,277 hectare. Of the other tahsils, Pharenda possess 2,673 hectares, Maharajganj 1,568 hectares and Basgaon 4,211 hectares. The groves chiefly consist of mango, guava, *jamun, aonla,* lemon *and mahua.*

FAUNA

The proximity of Nepal and the wide extent of forests are responsible for the presence of large number of wild animals in the district. The peculiar situation of the forests where they are nowhere more than 10 km. away from the village habitation makes the wild life prone to destruction. In the northern forests of Gorakhpur Forest Division tigers (*Panthera tigris*) are found in Nichlaul range. The population of tigers has of late increased in the division. Panthers (*Panthers pardur*) are found in every range of northern forests. Besides these *cheetal (Axis axis),* barking deer (*Muntiacus muntjak*), sambhar (*Cervusnicolor*), nilgai (*Boselaphus tragocamelus*) wild pig (*Sus secrofa*) and jackal (*Canis aureus)* are found in these forests.

In the southern forests of this division, which include Tilkonia, Banki, Pharenda and Pakri forest ranges, jackals, wild pigs, *nilgai, cheetal,* hares (*Lepus ruficandatus*) and squirrels (*Funamubuls palamrus*) etc. are found, while *cheetal* and pig are quite common in Pakri range, they are not so in Tilkonia range where the wild life is getting scarce. The wild life in Doma block across the Narayani canal consists of tigers, very few *cheetal,* para and *nilgai.* The other animals, which are found in these forests are jungle cat (*Felis chaus*), leopard cat (*Felis bengalensis*), monkey (*Macaca radiata*), and common red fox (*Valpes vulpes*).

Birds—The birds of the district are similar to those of the adjoining districts. The chief game birds found are of several varieties of ducks, partidges and pigeons. Besides these the other varieties of birds, which are commonly found in the district are peacok *(Pavo cristatus),* quail *(Coturnix couturix),* jungle fowl *(Gallus soneratti),* sinipe *(Cappella gallinago),* parrot *(Psittaculaeupatria),* kite *(Milivs migrans govinda),* crow *(Corvus splendens splendens),* vulture *(Gyps bengalensis),* bulbul *(Molpaster cafer), mynah (Acridotheres tristis), baya (Ploeeus philippinus philippinus)* or the weaver bird, sparrow *(Passer domesticus indicus), and bagula (Eqret alba modista)*

Reptiles— Different varieties of snakes and other reptiles are found everywhere in the district especially in the rural areas. Some snakes are deadly e.g. the cobra (*Naja naja*), *karait* (*Bengarus caeruleus*) and rat snake (*Ptyas mucesus*). The majority of snakes is non-poisonous, a few people die of snakebite almost every year. The other reptiles found in the district are crocodile (*Gavialis gangetieus)*, monitor lizard (*Varanus bengalensis*) and other varieties of lizard and python (*Python molurus*).

Fish— Fishes are found in the rivers, lakes and ponds of the district. The species which are commonly found in the district are *rohu (Labeo rohita), karounch (Labeo calbasu), bata (Labeo bata), khursa (Labeo gonious), bhakur (Catla catla), nain (Cirrihina miragala), raiya (Cirrihina reba), darhi (Barbus sarana), putia (Barrbus stigma), parhan (Wallagonia attu), tengra (Mystus aor),* *tingan (Mystus viatus), chegana (ophgicephalus straitus), girai (Ophicephalus gachuwa), patra (Notopeternus notopeternus), moi (Notoperernus chitala), mangur (Clarias mangur), singhi (Heteropneustes fossilis), chelwa (Chela bacila), belagara (Rita rita), and gonch (Bagarius bagarius).*

Game-Laws

The game-laws applicable to the district were governed by the Wild Birds and Animal Protection (U.P. Amendment) Act, 1934. It has been replaced by the Wild Life (Protection) Act, 1972 which made game-laws more stringent with a view to conserving wild life and preventing the extinction of certain species. The wolf, crocodile, gharial and peafowl have been declared protected species while certain restrictions have been placed on shooting of wild pigs, *nilgai* and some other species. The punishment for infringement of the laws has been made more deterrent.

CLIMATE

The district has a climate, which is more equable than that of the adjoining districts in the west and the north. The climate of the northern portions of the district is conditioned, to some extent, by the proximity of the hills in the north and the terai swamps. The year may be divided into four seasons. The cold season from mid November to February is followed by the summer season form March to mid June. The period form mid June to the end of September is the southwest monsoon season and October and the first half of November constitute the post-monsoon season.

The southwest monsoon usually arrives over the district by about the middle of June and withdraws by the end of September. The average annual rainfall of the district is 1393.1 mm. About 87 percent of the annual rainfall is recorded during the period June to September, July being the rainiest. The monsoon in the district generally advances from the southwest to the northeast. There is no large variation in the rainfall from year to year.

Temperature—The district has two meteorological observatories at Gorakhpur and Nautanwa. While the meteorological records at Nautanwa extend back to a few years only, those of Gorakhpur are available for a fairly long period. The data of Gorakhpur observatory may be taken as representative of the meteorological conditions in the district, except that the northern regions of the district have a comparatively milder summer as indicated by the records at Nautanwa. From mid November there is a rapid fall in temperature. January is the coldest month with the mean daily maximum temperature at 22.8 °C and the mean daily temperature at 9.3 °C. In association with cold waves in the wake of western disturbances passing eastwards in the winter season, temperature tends to go down to a degree or two above the freezing point. Day temperature begins to rise rapidly after February. May is the hottest with mean daily maximum temperature at 34.4 °C and the mean daily minimum at 25.1 °C. With the advent of the monsoon by about the middle of June there is appreciable drop in the day temperature. However, the nights continue to be warm. In September there is a slight increase again in the day temperature but the night temperature decreases after September. With the withdrawal of monsoon by the beginning of October it decreases progressively.

Humidity—During the monsoon and the post monsoon seasons the relative humidities are high ranging between 70 and 85 percent. In the winter months humidity decreases and in summer the air is comparatively drier.

Cloudiness—In the southwest monsoon season skies are heavily clouded or overcast. During the rest of the year clear or lightly clouded skies prevail, except in winter when the district is affected by passing western disturbances and the skies become heavily clouded or overcast for short spells of two or three days at a time.

Winds—Winds are generally light, slightly sharpening in the late summer and southwest monsoon months. Winds are mainly from the west in cold season. During the early part of the hot season easterlies begin to blow, but the westerlies predominate. Easterlies and north-easterlies prevail in late summer and monsoon seasons. In October, winds are light with large proportion of calms and the directions are mainly west, northeast or east.

Special Weather Phenomena—Occasional thunderstorms occur in late summer months. In association with spells of bad weather due to the passage of western disturbances, specially, in the latter part of the winter season, some thunderstorms accompanied with hail occasionally occur during the cold season.

Land and Forest : Use and Availability

In respect of land, it is important to note that India accounts for only 2.8 per cent of world's land and 1.6 per cent of the world's population (CSE, 1987). Whereas, district Gorakhpur of Uttar Pradesh accounts only for 0.00003042 per cent of the total area of the country. According to government reports, as much as 60 per cent of agricultural land is degraded to varying degrees in the district. Again about half of the total land, 1516.8 sq.km. to be exact, is affected by substantial soil erosion. According to soil scientists, soil is a non- renewable natural resource and it takes almost a thousand years to form a 2.3 cm thick layer from weathered rocks. The loss of 1mm of cultivable soil could cost 10 kg of nitrogen and 2Kg of phosphorus. What a colossal loss! Further more, our Common Property Resources (CPR) area that constitute the areas under the control of the community, has shrunk between 30-50 per cent over the past five decades. Gorakhpur accounts for 15 per cent of Uttar Pradesh livestock but only 0.5 per cent of pastoral land. Hardly 3.3 per cent of our total area is under grasslands while our domesticated animal population is close to 5 lakh in Gorakhpur. (Source: District Gazzetteer).

We have only 1 per cent of the total forest area of the country. Per capita availability of forests in India is 0.08 hectare as against the world's average of 0.8 hectare whereas in Gorakhpur, it is almost 2.8 hectares, which is so far a satisfactory figure. India's forest cover is barely 75.01 million hectare that is a mere 19.5 per cent of the total area against the required target of 33 per cent, which in the district Gorakhpur is still more than 37 per cent. According to a survey conducted by the Ministry of Environment and Forests, the country's forests have reduced by 5,482 sq. km. between 1995 and 1997, of which dense forests have been a major casualty. As against our rural population's annual requirement of 1.57 million tones of firewood, the production is only 0.58 million tones in the district Gorakhpur. The balance is met by illegal felling of trees by greedy commercial interests, for quarrying and construction, all in the name of development or crass consumerism.

Data relating to flora and fauna and animal life present an equally bleak picture. According to the Botanical and Zoological Survey of India, over 1500 plant and animal species are in the endangered category. Ironically, known as a mega diversity, India today has two of the world's 16 most threatened 'hot spot' locations (Swaminathan, 1994), out of which two are located in the district Gorakhpur :

1. Kusumhi Jungle Area and
2. Campierganj Area.

According to an estimate of the Gene Campaign and the Forum for Biotechnology and Food Security, the two voluntary organizations, India has diversity of 45,000 plant and 95,000 animal species. Of these, 90 percent plant species had already been patented by outside without least assertion of ownership by the Indian State, whereas there is no such record available or documentation made by the district administration of Gorakhpur. Even the district magistrate and the commissioner, Gorakhpur region were reluctant to speak on the issues of bio-diversity in their jurisdictional areas because of lack of information and non-compilation of the relevant data by forest department of the Gorakhpur district.

Water : Pollution and Problem

On the front of water resources, too, the scenario is quite depressing. On per capita basis, annual renewable water availability is 2170 cubic meters in India, as against the Asian average of 3370 cubic meters, which is pathetically low, 2.7 cubic meters in Gorakhpur. The annual availability of water per capita had declined from 5,236 cubic meters in 1951 to 2,257 cubic meters in 1991 in India, whereas it has further deteriorated in Gorakhpur to an extent of about 107 cubic meters, despite the very fact that the district has seen satisfactory monsoons for the last 7 years and is also endowed with natural rain fall because of its geographical location. Paradoxically, only about 40 per cent of the total water utilizable for irrigation is really used, while Gorakhpur district has perennial rivers – Rapti, Ghaghra and Kuana. Implying that, we are not yet able to fully utilize our irrigation to its potential level. Another disturbing trend is the massive shift of irrigation from surface water to groundwater. The proportion of cultivated area irrigated by ground water has risen from about one-third in 1965-66 to over 50 percent at present, resulting in a steady fall in the water table. It should, therefore come as no surprise that given the current water use practices, in future times it is water which will prove to be the effective constraint for production purposes, not land.

As for consumption, 300 million people in India have no access to safe drinking water while 700 million people have no access to elementary sanitation. The situation is further bleak in Gorakhpur district where more than 3 lakh people have no access to safe drinking water and have no basic health and hygiene facilities available to them. As many as 1,75,000 villagers are still without potable water (YCP, 1993). Even among the urban population, only about one-third of the people have access to safe drinking water, as per a survey report of Jal Sansthan, Gorakhpur, 2000-2001. One reason for it is that more than two-third of all drinking water is obtained from surface sources, seventy percent of which is polluted (YCP, 1993). According to a study entitled "Vision 2020" by the Washington based International Food Policy Research Institute (IFPRI), only 217 of 3,119 towns of India have partial (209) or full (18) sewage treatment facilities, thus, much of the sewage water remains largely untreated before being discharged into waterways and Gorakhpur is one such district in India having poor drainage and non functional sewage disposal plant. Thus much of the untreated sewage and other wastes are found to be scattered on the road sides or generally get mixed up with the municipal pipe lines rendering potable water unfit for consumption.

Scientists at the National Environment Engineering and Research Institute have concluded that 70 per cent of the available water in India is polluted. In the absence of proper sewage and water treatment facilities, thus, people across the country are consuming their own rubbish and excreta. Not surprisingly, according to a World Health Organization study, conducted almost a decade ago, 80 percent of the deaths in our country are caused by water-related diseases. Translated in economic terms, a World Bank study which evaluated the stupendous health costs arising out of the degradation of water, puts it at Rs. 19,950 crore a year. How then we can afford to have a healthy population that may bring sustainable development? As well as the amount that we spend on health and hygiene simply goes to waste.

Regarding this aspect, 'RASHTRIYA MALIN BASTI SUDHAR KARYAKRAM' has been initiated in the district headquarter of Gorakhpur on the instructions of Union Government for the year 2001-02. Finances worth Rs.165.84 lakh have been made available to the population of 5,05,566 living in slum areas of the district. The programme is mainly aimed at the construction and maintenance of roads, drainage system, community toilets and sewage disposal so as to provide sufficient ground for sustainable development. (Task Report of District Urban Development Authority, Gorakhpur on NSDP and Swarn Jayanti Urban Employment Scheme, 2001-02.)

From the above, it is clear that India in general and Gorakhpur district in particular, is facing an acute water crisis, with some estimates showing that almost its entire economic growth gets wiped out by the health costs of water pollution. Little wonder that the leading experts have warned that the world is heading towards "water shock" which might overtake the oil crisis. Ismail Seageldin, ex Vice-President of World Bank seems to be right in his observation; "many of the wars of this century were about oil, but the wars of the next century will be about water".

Air Pollution

The situation is equally grim in respect of air pollution both in the urban and rural areas of Gorakhpur district. Even at the national level, the three big metros in India, Delhi, Kolkata and Mumbai are among the 10 most polluted cities of the world. Over the last 25 years, air pollution has worsened due to industrial and vehicular emissions. Suspended Particulate Matter (SPM) levels have increased slightly while sulphur dioxide (SO_2) levels have declined. In the four metropolitan centers, suspended particulate matter is 360mg/cm compared to the WHO safety standards of 15. In District Gorakhpur it is estimated to be 138 mg/cm in densely populated urban areas. Add to it, the ever-increasing pollution load due to the emissions and vehicular exhaust, which is growing at a worrisome rate that has added to the already growing problem. A major source of concern is the exponential increase in petrol and diesel fuelled vehicles, the number of which on Indian roads has just doubled from about eleven million to twenty one million between 1986 and 1991. Their number is the highest in Delhi, which has on its roads more vehicles than all the other three metros (Mumbai, Chennai and Kolkata) put together. In Gorakhpur situation is no better because of the increasing number of two wheelers and diesel engine buses and taxies plying on the road. According to the Road Transport Department of Gorakhpur district, there has been immense increase in the registration of automobiles in the last ten years. In the rural areas, too, indoor air pollution due to smoke and fumes from the burning of biomass (wood and dung) that provides 87 percent of all cooking fuels is widespread (GOI, 1991). Nevertheless, the problem of pollution is more acute in urban areas than in rural. This is because over a third of Gorakhpur's urban population lives in slums. Also, about three quarters of urban households are without adequate sanitation. Aproximately 60 per cent of them have private toilets and more than 30 per cent resort to cpen defecation, thereby sullying the surrounding environs.

Coal combustion is another critical source of air pollution. While there is no escape from using coal, it pollutes the atmosphere by gaseous emission of Sulphur dioxide, Nitrogen oxide, etc. which is known to damage soil, vegetation and aquatic life of the region. In addition, it produces a tremendous amount of solid wastes, fly ash and bottom ash. The air pollution is taking a heavy toll of human health. This is evident from the fact that an average non-smoker in Gorakhpur is reported to inhale air pollution equal to half packet of cigarette a day. No wonder that 6 out of every 10 persons in Gorakhpur suffer from Asthma or some other breathing disorder. As many as 8000 of its inhabitants reportedly succumb every year to this menace, if a study of Gorakhpur Medical College could be believed.

Garbage

Fast increasing quantity and content of the garbage is another source of environmental pollution. Gorakhpur city alone, for example, generates about 10 lakh metric tonnes of city refuse every year. Though, of not quite the same volume as in the developed countries, the scale on which garbage is produced in this city range between 94 grams and 84 grams per capita per day. While the scale on which it is disposed of short falls between 103 grams and 164 grams per capita day. Thus, on an average, more than 30 per cent of the garbage remains unattended to. Furthermore, the character of the garbage has also undergone transformation. Unlike before, it now includes large quantities of plastics, metals, glasses, chemicals and medicinal and toxic wastes, which not only pose health hazards but also soil degradation.

SUSTAINABLE DEVELOPMENT : A CHALLENGE IN GORAKHPUR DISTRICT

No doubt, it is a herculean task to develop the District on model plan of sustainable development considering the worrying facts and figures mentioned above. The problem has got aggravated in the post independence period, thanks to our development endevour. Several explanatoray formulations have appearaed in this regard but the problem remains unaddressed. Therefore, if this area of eastern Uttar Pradesh has to play an important role in the development of industrial and commercial activities without harming the balance of environment and ecology, the government and administration must take immediate remedial measures to check further deterioration.

3

Role of Values in Promoting Sustainable Development

Twentieth Century was undoubtedly a century of unprecedented economic growth. Since 1900, the value of goods and services produced each year worldwide grew 20-folds, the use of energy 30-folds and the products of industry 50-folds. The growth accelerated after 1950, when President Harry Truman, in his address to the United States Congress, identified the largest area of the world as undeveloped. Declaring the new global policy, he remarked that all had to tread only one path and aspire to reach the same destination that was 'development'. The road to reach this was clear. Optimum production was the key to peace and prosperity. He also put forward a programme to save the people from misery and raise their standard of living through industrialization to be carried out with the help of U.S.' technical and financial aid.

Thus, the last 50 years have seen unprecedented growth and production in every field of human activity. In the field of energy, oil production was up by 567 per cent, natural gas by 1,143 per cent and nuclear power by 3,269 per cent. Humankind has reason to rejoice on these miraculous achievements. But now it is being realized that a heavy price was paid for this. Since then, 20 per cent of forests have disappeared and 20 per cent of land has gone out of production due to salinity, water logging and desertification. In India, 4.92 million hectares of forests were destroyed for expansion of agriculture, dams and other development projects between 1950 and 1980. Similarly, 36 per cent of the irrigated land area (20 million hectare, out of total 59 million hectare) was damaged due to salinity and water logging. The United States of America lost 745 million hectare of forests (93 per cent of its forest area) since Columbus stepped on the shores of America. The sad plight of tropical forests, which has become a matter of global concern, is well known.

Loss of Forest and Cropland

Noted environmentalist, Sri Sunder Lal Bahuguna[1] in one of his articles have given thoughtful account on the loss of forest and cropland, saying that in 1982 Food and Agricultural Organization (FAO) reported annual rate of loss of Tropical Forests as 11 million hectare a year, after 10 years it was 17 million hectare a year. Further, Temperate Forests in the rich countries have fallen prey to acid rains. In the words of Swedish forestry scientist Remrod, "our collective future is at stake."[2] Acid rains destroyed both the green cover and the purity of lake waters. In Sweden, 85,000 lakes have been affected. The horrors of radiation are well known. In 1990, world's 424 commencial reactors created some 9,500 tonnes of irradiated fuel, bringing the total accumulation of used fuel to 84,000 tonnes—twice as much as in 1985. Nuclear irradiation is also responsible for loss of cropland, "most of it in the Ukraine and Belarus. As a result of the 1986 Chernobyl explosion, at least half-a-million ha of cropland have been lost as communities no longer safe for humane habitation were abandoned".[3]

Symptoms of Perversion

The horrors of radiation, depletion of ozone layer, pollution and ecological crisis are the "gifts" of development, which have made man the butcher of Nature (*Prakriti*). The society, thus created is perverted (*Vikrit*). The three symptoms of perversion (*Vikriti*) are clearly visible in the modern developed society, which are as follows:

Rising Arms Expenditure

The first of these is the fear of war and internal insecurity. Military spending rose from 400 billion US dollars in 1959 to

24,545 billion (more than 60 times) in 1988. Even in 1991, when USSR was liquidated, it was US$ 19,165 billion[4]. To support their economies, the industrialized countries have to encourage the arms industry. Six countries-former USSR, USA, France, Britain, Germany, and China account for 90 per cent of global arms bazaar. Half-a-million best scientific brains are busy in devising the most powerful weapons of destruction. In 1988, per capita expenditure on war resources was $ 200[5]. It is in the interest of arms industry that war psychosis is maintained.

Areas rich in natural resources, specially in metals and minerals like oil, become the bone of contention. For instance, arms exports were propelled upwards primarily by the large-scale purchases of sophisticated weaponry by Middle East countries following the oil crisis in 1973 and 1979-80. Military import into the region surged almost five-fold from about $ 5.5 billion in 1972 to a peak of U.S. $ 26.8 billion[6] in 1984.

Internal insecurity within the countries, specially in poor countries, is a new problem created by the philosophy of development with economic growth. It creates desire in the individual to have more. Since desires can never be satisfied, so the individual is always in a state of discontent. War on terrorism is the collective manifestation of individual dissatisfaction. Politicians in western nations cleverly project it towards some others, *e.g. ex*-USSR or Saddam Hussain or Bin Laden, but in countries like India, towards other communities or rulers. Strangely enough, the most prosperous state of India, Punjab, has been the centre of terrorist activities. It is all because we have not given any significance to values in promoting sustainable development.

Depletion of Natural Resources and Pollution

Pollution and depletion of natural resources are other gifts of development. There can be no development till the chimney exits toxic gases and outlets poisonous effluents. Noise and dust also come with them. The trees disappear and the living soils die as the cities expand. India is losing almost 24000 sq.km. a year-i.e. nearly one per cent of its total land to non-farm purposes[7], whereas in Gorakhpur district alone out of 3483.8

sq.km of total land, approximately 500 sq.km. land area is affected by developmental projects, floods, soil erosion or non-farming activities. The other aspect of it is steep rise in carbon emission (36 per cent from 162 metric tonne in 1950 to 584 metric tonne in 1991)[8]. We have already discussed the loss of forests and cropland. Similar is the situation with two other renewable resources the cultivable lands and the rivers. The renewable resources, which sustain all life over our planet, have become non-renewable due to over-exploitation. Everywhere one can see examples of such follies being committed in the name of development. Water is the next victim after soil. This precious resource, which is rapidly diminishing, is not only being squandered by damming the rivers, but also by digging deeper and deeper to pump underground water.

Rising Hunger and Poverty

Poverty and hunger are the other gifts of development. It sounds strange that there should be poverty in a prosperous world. Out of over six billion people living on the globe, one-fifth lives under inhuman conditions. They never get enough to eat and have no access to safe drinking water. Development has brought prosperity to a few. The poor countries are drifting towards a debt-trap. They are under a debt of $ 1,35,100 crore (1.35trillion)[9]. In order to pay this debt, they export the fertility of their soil. The creditor countries now decide the land-use and international funding agencies, like the World Bank and International Monetary Fund, decide the economic policies of these countries. The result is hunger, famine and poverty, which many countries in Asia and Africa are facing. John Lupien, Director of FAO's Food Policy And Nutrition Division, commenting upon the progress of food production, has presented the statistics of hunger in these words, "Average calorie intake per person had risen from 2,290 calories per day 30 years ago to 2,700 calories a day in 1990. Despite this advance, the number of chronically undernourished in Africa had grown from 101 million to 168 million in the past two decades.... Nearly 13 million children under the age of five die every year as a result of hunger, malnutrition or related diseases and about

2000 million people suffer from a lack of vital micro-nutrients, such as iron, iodine and vitamin A".[10]

In India, we are heading towards the African situation. Already 26 per cent of the population lives below poverty line. The modest food of the poor is *dal* (lintels) and *roti* (bread). The prices of lintels have tripled in last three years and the prices of food grains are soaring. Coarse grains, like corn and poor man's vegetable potatoes will become too costly for the common people with the promotion of food processing industry. Since the religion of economics has made 'dollar' the God, earning of dollars has become the sole objective of most of the governments of South-East Asian African and Latin American countries. Instead of growing food for their hungry millions, they encourage cash crops, like sugar, cotton, tobacco, tea, coffee, etc. for export. This, in a way, is the export of the fertility of the soil and the moisture –the two most precious resources for survival of all beings.

To illustrate the benefits of development, statistics in per capita income are quoted. In India, per capita income in 1980-81 was Rs. 1,630 and in 1990-91 it rose to Rs. 2,227 (based on 1980-81 prices), but the rich gained from this as 20 per cent of the top share 50 per cent of the national income, while 20 per cent of the bottom got only seven per cent.[11]

Need For Harmony With Nature

We have seen that development with sole objective of economic growth which, of course, is the God of ruling elite even today, did neither bring peace on earth nor prosperity. Humankind has been living in a state of delirium in which destruction and disaster created by obliteration of nature were overlooked and economic gains-quantity are highlighted. In order to achieve this, we have lost the quality. The objective of a truly developed society should be the achievement of permanent peace, happiness and fulfilment. The first social revolutionary to challenge the false values and awaken humankind in this respect was Gautam Buddha. Buddha, in his quest to alleviate the misery, left the royal palace, became a common man, experienced all sorts of hardships, he under took fast for 40 days, after which light dawned upon him. He found that the root cause of misery was 'desire' (*Trishna*) and the way to get rid of this was ending the desires (*Trishna Kehhaya*). He differentiated between needs and desires. We should fulfil our needs, but should not run after desires. The prevailing concept of development encourages consumerism and keeps the individual in a state of perpetual dissatisfaction. That is the cause of all ills. Gandhi, in our times, also arrived at the same conclusion, when he said, "Earth provides enough to satisfy every man's need but not every man's greed".[12]

Two Big Lies about Nature

To reach a sound conclusion, we shall have to discard two big lies on which modern materialistic civilization, which flourished after Industrial Revolution in Europe, is based. These are;

1. Nature is a commodity; and
2. Society is only of the human beings.

Nature is not a commodity, but we ourselves, including all creations, are a part of Nature. Society cannot be of human beings alone as all life is interdependent. Thus, in a truly developed society, our interaction with Nature should be such in which Nature fulfils the basic needs of all beings. As human beings are endowed with a creative mind, it is our primary duty to sublimate nature with the help of science and technology so that she does not lose her capacity to sustain all life on the earth. The basic needs of all beings should be fulfiled from their surroundings with least efforts. These needs are oxygen, water, food, and shelter. These are common to all. Any system, which devoids living beings from these, cannot be tolerated in the name of development. As per the Hindu philosophy 5 debts (*Rina*)—*Dev Rina, Pitra Rina, Matra Rina, Rishi Rina and Bhoot Rina* are in harmony with the concept of sustainable development.

But, how to bring it down to Earth? Even the exponents of present system had to admit that Earth cannot bear the burden of affluence. So, austerity should be the first step. Thrift in the use of resources, both renewable and non-renewable is most essential. The present system, which relies more and more upon the use of non-renewable resources like metals and minerals, cannot survive because the supply of these is limited. The ancient cultures survived, because they sustained on renewable resources. While talking to a group of activist philosophers, Ivan Illysich was asked what was the way out of crisis of the civilization? He replied, "Perhaps the Eskimos can show us the way"[13]. Edward Goldsmith, on the basis of his study conducted for 32 years, also reached the same conclusion, when he says, "Inspirations must come from the world view of vernacular societies, in particular the ethnic worldview of the earliest period when people everywhere lived in harmony with the natural world".[14] These societies regard the living world or the biosphere, the basic source of all benefits and hence of all wealth, but this will only dispense these benefits to us if we preserve its critical order. The behavioural pattern of an ecological society must be to preserve the critical order of natural world for the cosmos.

Need to Find Alternatives

Modern life has become so complicated that we cannot do without certain things. So the next step should be to find out *Alternatives*. There is urgent need to replace the present energy system, which is inhuman, hazardous and temporary. The energy priorities should be: bio, solar, wind, tidal, geothermal and hydel from the run of the river.

Among some hopeful indications is the growing popularity of bicycle, which is the vehicle of cultured and scientific man because it does not emit polluted air, makes no noise and uses only 22 calories of energy to cover a distance of 1 km as compared to 1,153 calories by a single occupant of a car. Bicycle production has gone up from 20 million a year in 1960 to 60 million in 1990. Today the world bicycle fleet totals some 850 millions.[15]

Afforestation as a Remedy

Most of the problems faced by humankind may be solved by afforestation. This is essential to heal the wounds of bleeding

Mother Earth, to halt desertification, water-logging, salinity and soil-erosion from which the croplands are suffering. There are now very few forests, whatever we see today are the timber mines-monoculture plantations, which do not build the soil and conserve water. Soil and water are the two basic capitals, upon which all development depends. The remaining natural forests should be protected from the axe. Monoculture tree stands should be converted into mixed stands of trees, bushes, climbers and timber, which besides protection and improving the soil may produce food, fodder, fuel, fertilizer and fibre on a sustainable basis. This land-use will ensure availability of oxygen, pure water, food and shelter from the surrounding areas. Perhaps keeping this in mind, a special drive has been initiated under District Urban Development Authority (DUDA), Gorakhpur in slum areas to provide clean, fresh and healthy air to those people living in the most unhygienic and suffocated environment. For this purpose 120 tree-saplings are to be sown in every slum area and for their upkeep Rs.71400/- has been sanctioned for the financial year of 2001-02.

Gandhiji could foresee the potential of tree farming as early as 1926, when he remarked tree farming requires less labour and less water. Trees attract clouds and bring rains. He would judge every activity from ethical point of view and he was convinced that food from tree products would take people nearer to *Ahimsa* (non-violence). Now the scientists have found that trees produce five to ten times more on the same area of land as compared to cereals.

REALISING TRUE DEVELOPMENT

The two basic characteristics of development are:

1. It should be sustainable, meaning thereby the benefits enjoyed by this generation should also be enjoyed by future generations and that is possible only in a society relying more and more on renewable resources and

2. It should be ethical, *i.e.*, whatever is useful to one should not harm other beings. The present system is based on the wanton destruction of nature, in which a large

number of species become extinct. So far the human beings are concerned, "the gap between the rich and the poor is widening. The world has 200 billionaires and perhaps 300 million around the globe are homeless, living on sidewalks, in garbage dumps, and under bridges."[16]

One of the main causes of increasing disparities is the centralized system of production. It creates an army of unproductive people-managers, bankers, brokers, advertisers and transporters. The society, and ultimately the nature, bears the burden of these. Production of essential commodities should be decentralized. It will help in lessening pollution and exploitation.

The real development will take the society from *Vikriti* (perversion) to *Sanskriti* (culture). It is not going back, but going in the right direction.

Who Will Do It?

We have a blind faith in the governments and leave everything to those who sit on the citadels of authority. Governments cannot do this, because all establishments are static and they have only two weapons of fear and greed, which cannot bring a change. The change can come through people's movements, which should be initiated by humanitarian scientists (knowledge), social activists (action) and compassionate literary men, artists and journalists (devotion). In our philosophy of successful implementation of a noble act (*Yagyan*), the combined efforts of knowledge (*Gyan*), action (*Karma*) and devotion (*Bhakti*) are essential.

Acharya Vinoba Bhave, the walking saint of India, was also of the opinion that five powers (*Shaktis*)-good people (*Sajjan*), money power (*Mahajan*), learned people (*Vidwadjan*) government (*Rajya shakti*), and the common people (*Jan Shakti*)-should come together. It will be difficult to bring the vested interests in the fold for the change, but small groups of above three should form their small local groups, take up local problems, present their solutions and educate the masses.

Though they will be in a small minority, but, as Arnold Toyas bee, concluding his study of the world history, said, "During the course of history, changes were brought by a small minority." He quotes Buddha, Jesus and Gandhi to support his statement. This small minority is a creative minority, which presents workable solutions to the problems of humankind. The time for that is ripe, because the existing systems are collapsing.[17]

RELATIONS OF INEQUALITY

But why are we in the midst of an ecological crisis? Is it because of the so-called population explosion? No, countries and regions with low population densities are often far more ecologically damaged than those with high densities. An average American consumes 40 times more resources than an average Somali. A similar ratio is probably seen in the Indian context, if one compares the resources consumption of upper class dweller with that of an average villager. The American population with its profligate consumerism is far more destructive of the earth's environment than a much larger Asian population; similarly, the richest five per cent of Indian society probably cause much more ecological damage than the poorest 25 per cent.[18] Clearly, it is not mere numbers which are critical, but rather *who* consumes *what* and *how much*? This, in turn, is dependent on relations of power, on *who* decides what is to be done with an area's natural resources.

Fundamentally, it is these relations of inequality, which catalyse environmental destruction. Three main forms of inequity can be distinguished in this context.

Intra-generational Inequality

Hierarchical relations between classes, castes, races, communities, countries, and the sexes within one generation, can create conditions for environmental destruction. Those in power are able to dictate use of natural resources to suit their own needs, regardless of its consequences on others. For instance, the 'nationalisation' and take-over of India's forests by the British in 19th century was primarily meant to establish control over valuable resources. It was then that large-scale deforestation began, mainly to fuel the needs of British industry and the colonial expansion. In the process, those living closest to forests were systematically alienated. Unable to take decisions regarding their own local resources, they have gradually lost their incentive to protect forests; increasingly deprived of their sources of livelihood due to the demands of the urban-industrial sector, they have been forced to plunder the remaining forests to earn a pittance. And so is built-up a situation in which both rich and poor, both the powerful and the powerless, indulge in environmental destruction.

Inter-generational Inequality

Entire generations of human beings as yet unborn, who will be depending on the same natural resources we are currently using, have no voice in decisions taken regarding these resources. Humanity is clearly over-exploiting the earth's gifts-water, top-soil, minerals, fossil fuels eating into their 'stocks' at a rate which is likely to leave little for future generations. What we are likely to leave for them in massive quantities, on the other hand, are toxic wastes, barren wastelands, polluted waterways, and a handful of pest species, which have thrived on human wastes. If future human citizens have had some way of protesting, they would have a lot to say about our selfishness and greed. Sometimes, in a flight of fantasy, I have vision of deformed, starving, desperate children marching through the streets of the world's capitals, demanding justice children not yet born, but already condemned to pesticide induced genetic defects, the total loss of foodproducing top-soil, and the pollution of every drop of fresh water.

Inter-species Inequality

Humanity shares earth with a mind-boggling diversity of life-forms, perhaps upto 30 million plants, animals, and microbial species. This extraordinary explosion of biological diversity is not merely a source of wonder, but as fundamental bedrock of human existence as air, water and land. The world's agriculture remains heavily dependent on wide relatives of crop plants for continuous upgradation of cultivated varieties; over three-fourth of humanity depends on wild plants and animals for medicinal and nutritional needs; and our most basic aesthetic and cultural desires are fulfilled by other species. At another plane altogether, the fundamental right of other creatures to live on earth has also been recognized by various human faiths and cultures, and must increasingly become as much a part of our consciousness as the recognition of the rights of fellow human beings. Put simply, if I value my life and would not want others to take it away, I should be prepared to respect the lives of others, including that of non-human creatures.

Yet, ignoring all of the above, humanity has messed up the lives of countless other creatures in ways that bring no credit to evolution's most intelligent species. Simply because we have the might, we have considered it our right to colonise most of the earth's land surface, and increasingly even its aquatic habitats. Never before in evolutionary history has one species achieved such destructive dominance; over one fourth of all plants and animal species are condemned to extinction within the next few decades due to human hand. By the time you go through the motions of a day's life, at least one and possibly 25 species would have died out, never to be reborn. The inequality between humans and other species is the most pervasive of all inequalities, and its consequences the most tragic.

WHAT IS SUSTAINABLE DEVELOPMENT?

The notion of development too, needs closer examination. It has been forgotten that the term is the opposite of 'envelopment' and signifies an 'opening out', an expansion of possibilities. I would think, therefore, that development would include not merely material growth, but also, and perhaps more importantly, the blossoming of cultural, intellectual, spiritual, and other non-material value based aspects of human existence. Unfortunately, the Western notion of development has sidelined all of these aspects, and brought materialism to the centre-stage. Even intellectual and cultural growth today seems to be geared towards the aim of making money and accumulating goods. The contradictions and limitations of such an approach are very well typified by that seminal work, *Our*

Common Future (in 1987). An outcome of the World Commission on Environment and Development, and known also as the Brundtland Report, this book is increasingly being adapted by countries in national planning. In identifying many of the world's ills while analyzing the deteriorating environment, the report took a major step forward.

The Bulldozer of Homogenisation

One of the major flaws in the Brundtland Report is that it does not challenge modern development's most pervasive effect, the homogenization of biological and cultural systems. The notion of linear material growth does not respect the incredible biological and cultural diversity that the earth has harboured till not so long ago.

Conventional economic activities have been enormously destructive of diversity. Take for instance, forestry practices in India. Starting with the 'scientific forestry' of the British, mixed stands of vegetation were replaced by single-species plantations in an effort called 'improvement'. While enormously successful in enhancing the financial value of the resources, the concomitant destruction of biodiversity was great. The problem is that at least theoretically, it may be possible to continue such 'improvement forestry' indefinitely. Indeed, the relationship between biodiversity and sustainable development (as currently defined) is not at all clear; in other words, there appears to be nothing inherent in the notion of sustainable development that is biodiversity-conserving.

The same is true for cultural diversity. Western modernism with all its material trappings is sweeping the globe, road-rolling a multitude of cultures into oblivion. With that is also lost a wealth of beliefs, knowledge, and skills relevant to the wise use of natural resources: the intimate relationship of traditional farmers with their land, the acute tuning of the senses of forest-dwelling tribals to their surroundings, the uncanny intuition of fisher-folk communities to the movements and surges of acquatic fauna, the spiritual integration of conservationist ethics into economic practice by many traditional societies. Yet again, there appears little to suggest that the concept of sustainable development necessarily incorporates the conservation of cultural diversity, for it is possible to argue that development (read: economic growth) can be continued indefinitely in a culturally homogenous world.

Sustainable development, as currently defined, does not adequately address the perpetuation of intra-generational and inter-species inequity, and is, therefore, not acceptable from the point of view of the larger human goals of happiness, equality, justice, and peace.

If, as a species, we are truly attempting to live peacefully and wisely, then we will have to learn to respect life and diversity much more than we have in the last couple of hundred years. Biological and cultural diversity must become the keystones of our endeavour. As such, after fulfilling the basic human material needs of food, water and air, we must ask ourselves; *What more can we obtain for ourselves without violating the rights of other humans and non-human creatures? And if we get an answer that our material desires are impinging on the rights of other human beings and other creatures, then it is time to call a halt to the process of economic growth.*

THE RIO DECLARATION

The Rio Declaration on Environment and Development in 1992 puts people above all other concerns. It embodies several specific decisions reached for the first time at such a high level and on such a global scale. These include decisions to:

1. Establish poverty alleviation and full participation by all sections of the population, especially women, youth and children, as the chief guiding principles for global economic decisions.
2. Enshrine the sovereign right of each country to exploit its own resources in ways that do not cause environmental damage beyond its border.
3. Apply the principle of generational equity: that there should be equity between the rights and needs of the current generation and those of generations yet to come.

The present generation cannot want only expropriate our earthly habitat. It has a duty towards future generations to pursue sustainable development so that the planet might remain capable of sustaining those yet to come.

4. Observe differentiated degrees of responsibility among countries for environmental protection and sustainable development, although these are the common responsibilities of the world community. Thus, countries that pollute more have accepted that they should do more and pay more for environmental protection and sustainable development than others.
5. Bring recognition of the trans-boundary liability principle, under which a country, is held liable for environmental damage outside its borders caused by a polluter based on its territory.
6. Emphasize that countries should take into account the approach that the polluter should, in principle, bear the cost of pollution.
7. Insist that lack of conclusive scientific evidence should not be a reason for postponing urgent measures to prevent environmental degradation.

Therefore, a very pertinent question: is value based sustainable development possible? Does it fulfill the goals currently put topmost on the human agenda: equality, happiness, justice, peace? Or is it a contradiction in terms, with development being inherently unsustainable? As far as values in promoting sustainable development are concerned proper relationship between material progress and moral or spiritual values cannot be ignored. The issue of sustainable development, therefore, involves not merely the viability of industrial development or the degeneration or damage caused to the environment but also the normative evaluation of the moral basis of human life and civilization. The issue of contemporary crisis has been framed in terms of relationship between development and environment, but in fact, it extends to the vision of a human life routed in co-operation and contentment, peace and non-violence, and in harmony between man and man, and between man and nature. Even in the Gandhian view, economy is intrinsically inseparable from ethics. Underlying these facts, a set of values–spiritual or sacred–is found to be embedded with the concept of sustainable development. The modern industrial urban civilization makes it difficult to cultivate the values and virtues of civility–personal initiative and responsibility, concern for neighbourhood and community, and accountability for public conduct. What therefore is required for sustainable development is promotion and sustenance of values. We need creative thinkers and moral leadership that also provides political vision–religious divines who have strong routes in their religion and culture, but are humanist enough to incorporate the moral and spiritual approach for balanced and sustainable civilization.

NOTES

1. Bhauguna Sunder Lal, *Development Redefined*, The Indian Journal of Public Administration, July-September, 1993, *Vol.XXXIX No.3.*
2. Remrod, *Acid Rains in Europe*, IX World Forestry Congress, Mexico, 1958.
3. Zhores Medvou, *The Legacy of Chernobyl*, 1990.
4. ACDA, *World Military Expenditures and Arms Transfers* 1990, quoted by World Watch Institute (WWI), *Vital Signs*, 1992, New York, W.W. Norton & Co.
5. *Ibid.*
6. *Ibid.*
7. District Pollution Control Board Report 2000-2001.
8. *Ibid.*
9. Alan Thein Duming, "Income Distribution Worsening", in *Vital Sings, op. cit.*, 1992.
10. FAO, Fighting Hunger, Rome, FAO, 1993.
11. Third World Guide.
12. Pyarelal, *Towards New Horizons*, p. 12.
13. Edward Goldsmith, *The Way*, 1992.
14. *Ibid.*

15. Marcia D. Lowe, *Vital Signs*, 1992, p.72.
16. Alan Thein Duming, "Income Distribution Worsening", in *Vital Sings, op. cit.*, 1992.
17. Bhauguna Sunder Lal, *Development Redefined,* The Indian Journal of Public Administration, July-September, 1993, *Vol. XXXIX, No.3.*
18. Kothari Ashish, *Is Sustainable Development Desirable and Possible,* The Indian Journal of Public Administration, July-September, 1993, *Vol. XXXIX, No.3.*

4

Towards A Sustainable Society

What does sustainable development of society mean? Does it mean that the model of development be such that it does not excessively tax the energy or the patience of society for which it is meant? If that is what is implied, does not it then logically follow that development must be tailored to suit the changing moods and fluctuating tempers of society? In other words, development must aim at bringing about necessary changes in society, to be sure. However, its pace and impact must not be such that these agitate society too much or too deeply. If so, should not this brand of development give up its pretension to change society to correspond to the substantive vision it projects to be desirable? That is to say, in this perspective, development must step down from its status as a conscious, deliberate choice for reshaping conditions of human existence according to a priority defined vision of life and assume only a responsive role changing its character according to the changing mood of society.

In this sense, the concept of development refers at once to three situations. First, it refers to deeply felt discontent and unease about the patterns of development in the past. The sense of discontent and unease covers a variety of perceptions and realities ranging from pervasive poverty and its malignant consequences from human existence and dignity, on the one hand, to material affluence leading to loss of freedom and community, on the other. Second, it expresses and articulates the belief that present is malleable and the future determinable, that mankind is capable of escaping the pernicious effects of various patterns of mal-development whose roots go back in the past, and that the foundation of a better future can be successfully laid in the present. And lastly, it points to certain references and properties of the concept of development which, when translated into practice, would, it is believed, induce and sustain 'proper' development.

In view of these characteristics of the concept of development, it is simply inconceivable that development can be pulled down from its status as *idée directrice* (directive idea) and dressed up as an attractive toy for society to play with and discard when its fancy acquires a new attraction. Development, as understood contemporarily, constitutes a set of ideas to guide the endeavour of those who occupy ruling positions in society and are given the responsibility and authority to give a definite direction to society as it grows. The sustainability of development cannot thus be explored in terms of its compatibility with society.

It can be argued that every society has, at its core, a central organizing principle that gets progressively actualised in its institutions and cultural life in the process of social development. This organizing principle informs about their cultural as well as material life. Given this central tendency of growth implanted in all societies, it is only the question of identifying this tendency and then devising a mode of development that not only confirms to it but also becomes its accomplice in directing development of society in a way that makes progressive realization of this tendency possible. Such a development will eminently suit the genius of society and will, therefore, be quite sustainable because it is compatible with the basic 'nature' of society.

That is why a case study of Gorakhpur district has been made from different angles, so as to find out how far it has achieved social and economic sustainable development. All this can be seen in the following paragraphs as such.

LAND RECLAMATION AND UTILISATION

The district has a geographical area of 3483.8 sq.km. out of which the land has been put to various usages as follows.

Cultivated Area

During the period between 1856 and 1871, average area under tillage was 6,16,828 hectares or 62.9 per cent of the total of the district, excluding the large extent of forests. Appreciable progress in cultivation was recorded in subsequent years. At the time of the Settlement carried out between 1856 and 1871 it was found that 1,01,171 hectares of land was reclaimed in Mahrajganj and Gorakhpur tehsils of the district. The average cultivated area for the four years ending with 1887-88 was recorded to be 7,37,337 hectares. In the following decade the average rose to 7,95,618 hectares and this would have been even higher but for the temporary decline due to a cycle of bad agricultural seasons in the second half of the decade. The recovery was, however, rapid and in the decade 1898-99 to 1907-08 cultivation reached a higher figure than ever before, the average for the ten years being 8,26,549 hectares, while in the second half of the decade it was 8,35,292 hectares or 71.17 per cent of the total area of the district. In 1907-08, cultivated area on record was or 8,44,503 hectares. At present, after the district has been geographically distributed in the newly formed districts of Maharajganj, Siddharthnagar, Sant Kabir Nagar and Kushinagar, it still has 62.8 per cent of the total area under cultivation.

Culturable Land

Land under this category in the district includes wastelands, forests, groves, new or old fallow lands and also land otherwise classified as waste due to sandiness, barrenness, *reh* infection, etc.

The culturable land in the year 1970-71 included 55,628 hectares under forests, 16,688 hectares under groves, 11,334 hectares under culturable waste, 467 hectares under pastures and grazing grounds beside the fallow land, which measured 16,604 hectares. In the same year barren and unculturable land in the district measured 4,462 hectares. Besides, the total area of the land in the district which was covered under water, occupied by buildings and habitation sites, roads, railways, etc., was 52,058 hectares. But today the area of wasteland has increased specially because of construction activities and brick industry. Moreover, increasing population pressure has also affected the quality of land. Sewage disposal and waste management are pathetic and civic authorities of the district have failed to deliver goods. All this has generated ecological and environmental problem rendering development unsustainable.

Precarious Tracts

The precarious tracts of the district are few and not extensive. The main rivers of the district are Rapti and Ghaghra. Besides, Rohini, Ami and Kuwano Nadi flow through this district. The slope of land is from north to south. The southern portion of the district is low-lying and there are numerous small rivulets and *nalas* in this part. As a result, this part is usually most affected by floods and water-logging in rainy season. Ghaghra's fury is almost an annual calamitous feature. Vast areas and villages get submerged resulting in colossal loss to life and property and cultivation is worst affected. Precariousness assumes yet another shape in the tracts situated close to the banks of the rivers. In these areas the soil is usually sandy. An early cessation of rains plays havoc with the cultivation, the sandy nature of the soil precluding the construction of *kutcha* wells by the cultivators.

Irrigation

There are extensive irrigation facilities in the district. Little irrigation is needed in the *Kachhar* except in very abnormal seasons, for the natural moisture in the soil is oridinarily sufficient for the needs of the *rabi* crops. Again in the great rice tract of the north, where little attention is paid to the scanty *rabi*, irrigation is not needed except for the late rice, which is watered by means of channels from the *tarai* streams.

The returns of 1869-70 show an irrigated area of 3,67,225 hectares or 59.5 per cent of the net cultivation. Owing to the variation in the requirements of different seasons quoting only the figures of any single year may be misleading and consequently a fair idea of the irrigated area can be derived only from the averages of a more or less prolonged period. For the four years ending with 1887-88 the actually irrigated area averaged 2,09,581 hectares or 28.42 per cent of the total cultivation and the corresponding figures for the ensuing decade were 2,06,457 hectares and 29.51 per cent. From 1898-99 to 1907-08 the average was 2,65,667 hectares or 32.14 per cent of the area under the plough, the maximum being 2,98,065 hectares or 37.24 per cent in 1900-01. The general average was considered very high. The proportion of irrigation and cultivated land varied in different parts of the district.[1]

The per centage of irrigated area to the total sown area was 36.5 in 1971-72 (whereas the state average was about 40 per cent) which increased to almost 85 per cent by 2001 to 2002 of the total cultivatable land.

Means of Irrigation

For irrigation purpose the district has a network of canals, tube-wells and minor irrigation works like *pucca* wells fitted with pumping sets and Persian wheels. By 2000-01, the cultivated area under irrigation had increased to almost 85 per cent through network of canals and tube-wells. The popular means of irrigation in the district are:

Canals

The numerous rivers have been incessantly bringing havoc in this area in the shape of floods and waterlogging in abnormally wet years. The wayward monsoon also results in droughts and worst famines in the area. Wells and tanks were too insufficient and often of no use in years of acute drought. Canals from the perennial rivers were considered the only hope to solve the problems and to guard the people against the vagaries of monsoon, droughts and floods. As early as the beginning of 20th century the Tharus adopted a system of canals for the *tarai* streams of Binayakpur and Tilpur parganas the water being collected by means of earthen embankments and conducted to fields along channels called *Kulas*. In 1907 it was proposed to construct a gravity canal from the right bank of river Great Gandak near Bhaisalotan or Valmiki Nagar in Bihar. The scheme was, however, rejected by the State Government on the ground that the area produced mainly paddy which did not require any irrigation in normal years.

Food shortage in the country during and after Second World War brought the problem again to the fore. In 1947, the Central Government requested the government of Bihar State to investigate the possibilities of taking off canals from both sides of Great Gandak river for providing irrigation facilities in the districts of Gorakhpur and Deoria in Uttar Pradesh and also a large area in Bihar State. In 1954 a preliminary project report was prepared and submitted to the Government of India by Bihar. It envisaged an expenditure of Rs.3,194.0 lakhs[2].

As a measure of immediate relief, the construction of an inundation canal, namely *Naraini-Pokhra* canal was taken up by the U.P. Government in 1954. This canal takes off from the right bank of the Great Gandak at *Pathrawa Ghat*. The canal was completed in 1956.

The U.P. Government has also prepared a project plan for the portion of Western Gandak canal system covering part of Nepal and the two districts Gorakhpur and Deoria (now Maharajganj also) in this region envisaging a total cost of Rs.1,547.40 lacs. In 1961, the U.P. the Government submitted to Government of India, a revised plan for the works in the State entailing an expenditure of Rs.1,515.69 lakhs proposed to provide irrigation to an area of 2,84,089 hectares. Work on the Western Gandak canal project (for the area falling in U.P.) started in the same year whereby the existing Naraini canal system was merged with the new project. The remodeling work of Naraini canal was completed in 1969-70. The work on Gandak canal project was started in 1967-68 and completed in 1974.

The Gandak canal system has been designed for providing irrigation to the *Kharif* crops comprising nearly 40 per cent of the total *Kharif* area and 20 per cent of the total *Rabi* area in the district. The main western Gandak canal takes off from the right bank of river Gandak at Valmiki Nagar. For the first 19 kms. it flows in Nepal territory and thereafter it enters Uttar Pradesh and crosses the State boundary. The length of main canal in U.P. is about 112 km. It was run for the first time in December 1972. There are 17 canal systems, which take off from the main western Gandak canal from 19 kms. to 131 kms. The main systems are Deoria branch, Malhia branch, Khajuria branch, Chaff branch systems besides Rohini canal, Danda canal, Srinagar Tal, Naraini canal, Bakhira canal, Kuwano pump canal, Ramgarh (east and west) canal and Gandak canal. The total length of the canals is 2,388 km. in Uttar Pradesh. Therefore, it can be argued that with networking of canals, sufficient attention has been paid on agro-based economy of the district, which is in harmony with the concept of sustainable development.

Tube-wells

The average depth of underground water for the boring tube-wells is between 91 to 107 metres. The northern part being *tarai* has abundant rainfall. There are many swamps in these parts and the farmers use water of swamps for irrigation by installing pumping sets. Well-to-do farmers build their own tube-wells for irrigating their land, but the excessive use of tubewells is taking a heavy toll on the water-table in this area for which district administration is not taking any preventive measure.

Wells

The average depth at which water is found is about 4.5 metres below the surface in *bangar* and much less in the northern tracts. As the level of water is high in *bangar* the farmers build temporary wells at lesser cost. In 1972 there were 19,699 masonry wells which irrigated an area of 30,813 hectares. The shallow wells are worked by means of *dhenkli* or lever, the *mot* or leather bucket. But gradual lowering of water table has rendered many wells dry and inconsistent. More and more use of bore-wells has again depleted the water table. In the city and other densely populated areas, people use monoblock pumps which again add to the problem of water supply. If sufficient measures are not taken, the problem of water would be of acute shortage and potability in not far off future.

Tanks and Lakes

Tanks for irrigation are as common as in any other part of eastern districts. These are the usual square or rectangular excavations dug down to water level. Gorakhpur is remarkable for the number of its large perennial lakes formed in most cases in abandoned channels of rivers. Besides these, the district possesses a vast number of temporary swamps and *Jhils*. Most important among these are *Ramgarh Tal, Momingarh Tal, Naduar Tal, Amiar Tal, Bhenri Tal* and *Chillua Tal*. Lakes and *Jhils* are also used for irrigation. The water is conducted along narrow channels, to the field by the use of *beris* or saving-baskets of wickerwork.

Minor Irrigation Works

As the rainfall is not assured, drought conditions often threaten agricultural production. The canals and State tube-wells are as usual very unreliable, poor and insufficient suppliers of water to the cultivation. The government has taken up the programme of providing financial assistance to cultivators in shape of loans and grants for constructing private wells, installation of pumping sets and Persian wheels *(rahats)*, etc.

Land and Soils

While the greater part of the district of Gorakhpur falls in the eastern sector of the Indo-Gangetic plain whose two types of alluvial soils, (i) Older alluvium *(bangar)* and (ii) New alluvium *(Khadar)* are generally found, a broad strip in the north forms part of the tarai belt which runs parallel to *bhabhar*, along the foothills of Himalayas, from the west to the east.

In the central part of the district, the soils are highly calcareous. Soils in the tarai area have developed under sub-humid climate and high water table conditions and natural vegetation of tall grasses. The parent material is medium to loose textured alluvium. The soils are characterized by the presence of a dark coloured organic matter with rich surface horizon. Other tracts in the district possess soils, which are neutral to moderately alkaline and calcareous, and have sometimes, well developed clay accumulation horizon in the subsoils.

Bangar soil varying from loam to sandy loam, is distinguished by its poor lime content and is generally mildly alkaline to slightly acidic in reaction. A third type of soil called *Dhuh* occurs near river-banks, which is liable to inundation.

The Soil Survey Organization of the State carried out a soil survey of the district in 1951, and found the soils deficient in nitrogen and phosphorus content. The nutrient index of potassium was, however, not discouraging. In *tarai* the main staple crop is *late rice* (a type of delayed rice produce), for which pargana Binayakpur is famous. In the *bangar* the nature of the crops depends on the compostion of the soil. In the north, where the rainfall is abundant and throughout the clay areas, rice is chief staple, the early variety being grown on the higher levels and the late in depressions which hold sufficient water.

Harvests of the District Gorakhpur

There are the usual *Kharif* and *Rabi* harvests in the district, the *Zaid* being very insignificant. The *Kharif* or rainfed crops are sown in June and July and harvested in September or October, while *Rabi* or irrigated crops are sown in October or November and harvested in February or March. The crops harvested more than once in a year (*Dofasali*) covers a large area in the district. Principal *Kharif* crops include rice, maize, *arhar* and *bajra* while *Rabi* crops are wheat, barley, gram, pea and *masoor*.

Sugarcane is one of the important cashcrops of the district, especially in *tarai*. A considerable area is covered under cultivation of oil seeds such as linseed, tora, rapeseed and mustard. Of these the linseed is commonly sown as a second crop after rice while the rest are almost grown with various *Rabi* crops, especially cereals. Formerly the district was known for cultivation of poppy also, from which opium is extracted. In 1957 its cultivation was stopped totally.

After independence, it was realized that the traditional method of cultivation could not cope with the ever-increasing demand for food grains. Improvement and changes in the pattern and technique of cultivation were therefore adopted. Agricultural production can be increased either by extensive cultivation or by intensive cultivation. Extensive cultivation is possible by bringing larger areas under cultivation, by reclaiming wastelands, fallow land etc. Intensive cultivation consists of applying scientific methods in cultivation, provision of better seeds evolved through agricultural research and use of improved implement and chemical fertilizers etc. The 60s of the 20th century witnessed the green revolution in the country, under which scheme of intensive cultivation and sowing of high yielding seeds have been implemented. The government agricultural farm in the district and various other agencies of the Central and State Governments, the Food and Agricultural Organisation of the United Nations, Agricultural Colleges and Universities and Research Centres in the country evolved not only high yielding technology but also eco-friendly method of agriculture. But use of excessive chemicals led to deterioration of the quality of soil, thus badly affecting sustainable development.

The old indigenous implements and tools such as *Hasia*, *Phawra* and *Khurpi* etc. were replaced by modern devices using higher fuel consumption, which again aggravated the problem of environment.

Industries

The weavers of the district have been producing cloth since ancient times, and the textile industry had its own guild rules and regulations, which were recognized by the law of the land. Sometimes head of the guild was an important personage of the court. The other, less important industries of the district included ceramics, metal casting and forging, and footwear.

In the medieval period, with the emergence of the middlemen as financiers, the industrial guilds disappeared. Much of the returns from a bargain were pocketed by the middlemen leaving the artisan in a state of poverty. However, the nobility at times helped the artisans who could sell their goods at reasonable prices. Akbar had abolished many cesses and taxes, thus relieving the condition of the artisans. In this period the handloom industry prospered in the distrtict and a large number of Muslims joined the trade. The artisans lived in comparative peace till the annexation of the district by the East India Company, in 1803. The Company discouraged Indian industries, so that textiles and other goods manufactured in England could be sold in India. The weavers stopped producing fine drapery and only *Garha* (coarse cloth) was manufactured as noticed in the first decade of the twentieth century. *Dhasa,* mixed fabric of cotton and wool was woven in very limited quantities at Gorakhpur. In this period, a unique industry of embroidering leather was carried out by two or three families at Gorakhpur. Deer skin was tanned in rich reddish-brown with *sal* bark rendering the outer surface soft like velvet. The leather was then embroidered with coloured silk and finished with a gilt edge. Round pillow cases, embellished with rich works in Persian were then carved out in suitable sizes. Price of a piece ranged between 8 annas (50 paise) to a rupee. Chair and table covers and prayer carpets were also made out sometimes. The design was conventional but elegant. Metal vessels were also made at Gorakhpur.

The British encouraged the cultivation of indigo, which yielded a blue dye. By 1888, its plantation had no doubt, spread all over the district but the peasants prefered to grow sugarcane and opium. A factory was established at Babhnauli, to extract the dye whose production had to be stopped when a synthetic process was evolved to manufacture artificial ultramarine. Indigo cultivation was given up and the industry came to an end by the 30s of the 20th century.

Production of sugar has been an old industry of the district. The farmer used to boil strained sugarcane juice in open pans till it became thick enough to form crystals, drain out the molasses, pulverize the brown stuff, subject it to decolourisation and convert the anhydrous mass into refined sugar, adopting a method that was primitive by modern standards. The indigenous system went out of circulation in areas where the sugar mills employing a better technique to produce sugar, started coming up. By 1932, there were five sugar factories, at Sardarnagar, Pipraich, Ghughli, Siswa Bazar and Anandnagar. A Jute mill was established at Sahjanwa in 1935 and a distillery was opened at Sardarnagar in 1951. Thus, modernization and use of technology led to the environmental hazards. Earlier, though the technique deployed was primitive in nature, yet was environment friendly. But the new technique despite giving rise to higher production, proved detrimental for sustainable development as many of the artisans and workers became unemployed.

After 1947, a number of small-scale industries cropped up at Gorakhpur, which had a large railway workshop, but by and large, the industrial progress of the district has been rather slow. The census of 1961 indicated that the district had 3,568 workshops and factories, which fall below the State average of 4,460 per district. Of these 2,125 were located in the rural areas and 1,443 in the urban centers. However the district gained a significant place on the industrial map of the State with the opening of a fertilizer factory at Gorakhpur in 1969 (now it is non-functional). Another large-scale industry, the Saraya Steel Complex, Ltd., has been established at Sardarnagar in 1974 with an installed capacity of 10,000 tonnes of plain carbon steel ingots per year. This further aggravated the problem of sustainability in development, as the layout plan of these industries is not in consonance with the present environmental laws.

Power

The first powerhouse in the district was established in 1928, followed by another a few years after, both managed by the Jhansi Electricity Company. However, only 4.8 per cent of the establishments used power in 1961.

The U.P. State Electricity Board took over the management of powerhouses from the Jhansi Electricity Company, in 1948.

In 1969-70, the total consumption of power in the district was 26,68,52,387 units. The per centage consumption of energy in the various sectors of the economy in the year 2000-01 is given below:

Table 4.1

Sector	*Per centage consumption of power*
Industries	94.0
Agriculture	2.7
Domestic	2.2
Commerce	0.6
Others	0.5
Total	100.0

Source : District Gazzetteer.

It is evident from the table that agriculture sector of economy, which is the backbone of Gorakhpur district, utilizes negligibile per centage of power, rendering agronomy ineffective in the development of the district.

Heavy Industry

The Fertilizer Corporation of India Ltd. was established at Gorakhpur near the Nakha Jungle railway station in April, 1963. However it commenced the production of urea on January 1, 1969. The factory is owned and managed by the Fertilizer Corporation of India, Bombay (now Mumbai). For the last ten years it has become non-operational, leaving a number of its workers unemployed and harming socio-economic structure of the district. Not only this, the wastes coming out of this factory had not only deteriorated the soil content but also had polluted water and vegetation. No effective measure has been taken to prevent this contamination by the district administration of Gorakhpur.

Saraya Steel Camplex Ltd., Sardarnagar

This factory was established in 1974 with a total investment of Rs.53,10,000 and a capacity of 10,000 tonnes of plain carbon steel ingots per year. Steel is being manufactured from the iron-scrap available from the North Eastern Railway Workshop, Gorakhpur. The factory is equipped with an electric arc steel-melting furnace. Despite this, the contamination it causes to the environment is voluminous. Pollution Control Board of the district is almost non-existent and the factory has not received any clearance from the Department of Environment and Ecology in the last 3 decades. It clearly shows the apathy and callous attitude of the government officials towards sustainable development in the district.

Large-scale Industries

Sugar, gunnies, jute strings, oil and oil-cake, and alcohol are produced in large-scale units, which are located at Sardarnagar, Anandnagar (now in Maharajganj district), Siswa Bazar, Ghughli and Sahjanwa. These industries too, do not comply with any of the present environmental law. Despite this fact, no action has been taken against the management of these industries.

Sugar

The Saraya Sugar Mills Private Ltd. was established at Sardarnagar in 1919. The Punjab Sugar Mills Company Ltd. was established at Ghughli in 1920. The Mahabir Sugar Mills, Private Ltd. was established at Siswa Bazar in 1931. The Ganesh Sugar Mills, Anandnagar and the Diamond Sugar Mills Ltd. Pipraich were established in 1932. Most of these sugar factories have either become redundant or non-operational. Thus, large areas are under the cover of these factories making the use of landed property unprofitable and non-ecological. Neither the State Government nor district administration has taken any step to revive these industries, which otherwise can prove to be a major source of employment and economic sustenance.

What is distinctive about the dominant paradigm of development today is the recognition of the primacy of the value of augmentation of fortune in life. Rapid scientific and technological advances have kept alive the belief that it is possible for everybody to enjoy affluence if not today then, at least, the day after. It is because of the prevalence of this belief that the pursuit of self-interest has emerged as the focal point around which life activities get organized. It is true that the unbridled pursuit of self-interest leads to discord. But discord, as Kant asserts, awakens in man all his pent up powers and propels civilization on the path of ever-higher achievement. Needless to say that it is this conception of development that has been given effect to, since the industrial revolution. It cannot be doubted that this mode of development has improved the life conditions of millions of people all the world over. However, it is also true that this has also endangered the very existence of man by exacerbating ecological degeneration. The operation of the physical limits of growth has, therefore, led to the reluctant recognition of saving man's natural habitat from the ravages of man's greed.

Difficulties Involved

If one could have all the good things of life, as we are accustomed to defining them today without harming our environment, what can be better than this? But a moment's reflection suggests that it is quite a feat to laugh loudly and keep your mouth shut all at once. First of all, the cost of cleansing the already over-polluted environment and keeping it clean is so prohibitive that most poor state governments of the country cannot afford it. Second, the very success, if it comes our way, in doing so would push the demands of goods and services to such an extent that no political system can afford to either satisfy or ignore them. To try to satisfy them is to exploit nature much more than is the case now and, therefore, compound the ecological problem further; to ignore them is to commit political *harakiri* by pushing the level of people's dissatisfaction to a very high pitch.

Auto-Control and Sustainable Development

But it is inherent in the dominant paradigm of development that it erodes not only auto-control but also the worldview that makes auto-control possible. And it is quite difficult to imagine that without auto-control one can think of sustainable development. Thus, the criterion of compatibility cannot be derived from a flimsy hope that science and technology would eventually succeed in developing ways and means not only to clean the already polluted environment but also to devise a production process that does not yield any pollutant. If this is so, the criterion of sustainability must be looked for elsewhere.

It also needs to be emphasized here that such a criterion can be derived neither from the idea of individual welfare nor from that of collective good. It has to be sought somewhere in between, i.e., a perspective that reconciles satisfactorily the claims of the two orders of individual existence, his private and public life. Once it is realized that the problem of sustainability of development can be resolved only with reference to the urgency of this reconciliation, it will also be realized that the basis of this resolution is the recognition that man is an integral part of a larger order, that man is not the measure of everything.

Poverty in rural areas of Gorakhpur forces people to cultivate more marginal lands, which results in soil erosion, depletion of shallow water resources, and consequently lower crop yields, leading to greater poverty. Trying to survive on a daily basis, the poor have often little choice but to continue to overexploit their resources, which in turn also reduces the chances of their offsprings ever-escaping the poverty cycle and an unsustainable environment. Of the estimated thirty seven lakh eighty four thousand seven hundered and twenty (37,84,720) number of people, the poor people of the district, some four lakh fifty thousand (4,50,000) live in low-potential agricultural areas, a similar number live in highly ecologically vulnerable areas susceptible to soil erosion, land degradation, floods and other ecological disasters, and about one lakh are in urban slums.

The poor are often the victims of environmental stresses caused by the actions of the rich. They are forced to live in areas more vulnerable to natural disasters, industrial hazards and air and water pollutions. The growing migration from rural poverty to urban squalor also is of increasing concern as this concentration of poverty leads to the breakdown of the urban services and social systems, resulting in increased crime and a destabilising socio-political environment. The burden of poverty and deprivation also fall more heavily on women and children and certain minority groups. These target groups require specific measures in poverty alleviation programmes.

An environmental policy that focuses mainly on the conservation and protection of resources without due regard to the livelihood of those who depend on them not only will have an adverse impact on poverty but is also unlikely to succeed. Similarly, development policies which focus mainly on increasing production without concerning themselves with the long-term potential of the resources on which production is based will sooner or later run into problems of declining productivity and consequently greater poverty. An effective and sustainable development strategy to tackle the problems of poverty and environmental degradation should simultaneously focus on resources, production and people. Therefore, sustainability in the society of Gorakhpur still appears to be a distant dream and if this problem of sustainable development is not dealt with properly, it may turn into a chaos, which if not contained may bring the catastrophe.

NOTE

1. District Gazzetere.
2. Ibid.

5

Participatory Approach in Transfer of Technology For Sustainable Development

The Earth is a functioning whole. Its ecological integrity must be maintained; this includes the reduction of social and cultural vulnerability and biological diversity.

The real issue of social and environmental crisis is not one of protecting the environment *per se*, but of managing resources to attain sustainable social, economic, and physical development of equity with respect to gender, religion, and race at all levels of human organization, self-reliance, respect for biological and cultural diversity, participatory democracy, cultural and political integrity, and access to education and information leading to active co-operation amongst all people.

This issue has to be tackled by altering excessive consumption patterns as well as through the acknowledgement of citizens' right to economic and social improvement; the satisfaction of basic needs, material and non-material must be guaranteed and fulfilled for present and future generations.

Citizens, enterprises and governments must work together to protect global climate system on the basis of their common but differentiated capabilities and responsibilities for sustaining life on earth.

We have had in this country more than 50 years of 'planned development'. The broad goal was growth with social justice. Growth has always been measured in terms of the rate of growth of economy, of GDP. This growth of income and wealth was expected to trickle down, seep through, to the disadvantaged sections of society in aid of social justice. By 1970s, it was clear that not much social justice was being achieved by this method. The reason basically was to be found in the planning approach which was not directed specifically to the human factor, but rather to making more steel, heavy machines and so on, chasing levels of production and consumption of such items in the economy to reach self-sufficiency as well as developed–world norms which were supposed to reflect a desirable industrial base. In the public sector, financial and other resources were poured into these efforts, as a result the levels of consumption and living of the vast majority of people outside the organized industry and labour-world suffered because of inadequate attention and resource inputs.

COMMITMENT TO THE POOR

The growing concern about the condition of the poor led to official calculations of poverty line and poverty ratio from the 1970s. But there was no change in the planning approach for the industrial sector that continued to receive massive flows of public money, for investment as well as to subsidize losses in the public sector.

Later, more and more public money was found to sustain larger employment programmes in rural areas, which provided more purchasing power to rural people. Such programmes were the Drought Prone Area Programme, the Cash Scheme for Rural Employment (it was perhaps more of a political measure following the 1969-split in the Congress Party than anything else), the Food for Work Programme, the Half-a-Million Jobs Programme of the mid-1970s, and the Jawahar Rozgar Yojna of the 1980s, all of the Central Government. These were palliatives, with varying degrees of success in implementation. In the states, a notable initiative was the Employment Guarantee

Scheme of Maharashtra in the early 1970s, which, however, did not find favour with central Planning Commission for some years. So was the response of the Government, as the Commission felt that it was not in accord with 'national' priorities'! So also, the West Bengal's practice of the 1970s, of targeting welfare and works programme at specified small percentages of the population (assumed to be so poor as to need such aid) in stated months every year, identified particularly for each, ran into heavy weather in the Planning Commission, thus, rendering concept of sustainable development untenable. Unless and until downtrodden and underprivileged sections of society are taken care of, no technological, commercial and environmental development is viable. The 'trickle down theory', therefore, has not been able to gain much in the whole of Uttar Pradesh, particularly in Gorakhpur region. The other major causes for non sustainability can be enumerated in the following lines.

Tardy Land Reforms

Except in few states, which efficiently implemented land reforms and provided to vast numbers of the erstwhile tenancy their own land assets, land reform laws did not decisively open up access to land resources for the poor. Bihar and Uttar Pradesh are commonly accepted examples of indifferent implementation of land reforms. One would infer that at the political level, the commitment to land reforms which would transfer land from large holdings to number of tenants was not the same in all states. A very senior official, who had much to do with the implementation of land reforms in Uttar Pradesh, had observed (in a seminar in 1986) that much of the officialdom in the government departments concerned was inclined to the side of the *zamindars*. One recalls a minute of the then member of the Constituent Assembly from the district Gorakhpur, Shri Shibbon Lal Saxena, published in the 1970s, containing a number of articles on various aspects of development in Gorakhpur, to the effect that it was not easy to improve the lot of the tenancy by legislation because the judges in the courts were generally from the same social class as the *jenmis* and tended to favour the latter.

Even today, major cause of social tensions in the district is related to improper management of land records. The tenancy and ownership relations in the district have further vitiated the law and order problems. Rampant corruption in revenue and consolidation departments is a common practice. Any amount of developmental effort would be of no use unless large scale effective land reforms are implemented at the district and state levels.

Not Saving Resources for Social Services

What about political and official-level commitment to the poor and their betterment (apart from pursuit of GDP growth)? One finds more than one stream in India all along, directed to such 'social justice'. One stream consisted of legislation for land reforms, freeing bonded labour, outlawing untouchability, labour welfare laws, public distribution system, prescriptions for job reservations and so on. These have been put into effect with varying degrees of success in different parts of the country, and often spasmodically. I have referred to another stream, the different employment generation programme over the last 20 years or so. The third stream consisted of plan allocations for special programmes, such as Integrated Child Development Scheme (ICDS) and the Integrated Rural Development Programme (IRDP) and for the 'social services' that is, education, health, drinking water, etc. Often these services were pushed to the wall whenever there were not enough financial resources left in the plan after taking care of the sector of agriculture, industry, power, transportation and other economic services. For instance, in 1981 when the results of the Census of that year showed that the population of the country was larger than what had been projected by the '*pundits*', the consequences for primary education were that the achievement of universal primary education would slip by some years. The planners and the political executive accepted this with scarcely a ritual murmur of discomfort; the pre-emption of available resources by the economic services, etc., was not questioned even for a moment, nor was there any attempt to examine whether resources could be saved for the social services by greater efficiency of their utilization by those sectors. The view of the political executive, exemplified in this case, was reflected in the official levels also. There have been instances when chief secretary of a state government who readily cut plan provisions for public hospital beds to find more money for the industrial sector. That was in the 1960s, but the passing years made no large difference in this kind of approach in the governments.

People as Mere Statistics

One wonders what happens to politicians and officials when they move from district and state levels to the Central Government level. One would imagine, or hope, that they would be able to provide-to the Central Government at the political level, to the Planning Commission and the senior official machinery-the kind of feel and understanding of how things work on the ground, whatever the policy objectives, of what one may call needs and problems of the poor who otherwise can hardly make themselves heard. But it is an irony that most of them start speaking, soon enough, the language, which projects a 'Central' point of view to Parliament as well as to the states. I do not wish to enter into a debate here as to. why this transformation takes place, but I do believe that it is one of the important factors tending to distance Central Government's decision from the needs and realities on the ground, of the disadvantaged sections of the people, and also from what is administratively feasible or desirable.

Further away from the field one goes, particularly where planning and administration of social services are concerned-the more the tendency to think of people in terms of statistics, and that too for the country or the state as a whole-increases; it is also then easy to ignore them, say, when one wants to lease or acquire large areas of land, for example, for a steel plant or an irrigation project, or for defence purpose, or for mining. And, this is what happens. One sees vast, unused tracks with one or the other enterprise or government department in a country where land is a scarce resource overall, and even in area where there is not enough land to sustain local population. One such example can be given of the Planetarium being constructed in Gorakhpur, near Ram Ghar Tal Project, for the last two decades.

It is gross misuse of public funds and vast track of land. For today's population, the landmass of the country provides less than 0.38 hectare per capita compared to about 0.9 in 1951. Alternatives for using less land rarely receive consideration, while far too readily government departments acquire many times the area they immediately need, on the basis they might need more in the next 10, 20 or 30 years.

This can possibly be termed a colonial paradigm. The rulers and their supporters are too ready to exploit resources that could otherwise sustain local people in some manner or the other. One is in good company referring to the colonial pattern.

Neglect of Human Aspect

In this kind of an administrative and political milieu, I believe it is inevitable that development is often neglectful of human aspects. I believe also that the results of such development are unsustainable both in the language of the environmentalist, and in social terms. How else would one really explain widespread and increasingly strident 'Green Movement' in many parts of the country? How else would one explain the steadily increasing disaffection and disillusionment which have led to militancy or resistance movements against authorities spreading in many parts of the country variously called in Bihar, Andhra Pradesh, parts of Maharashtra, Uttranchal, the Northeast and Uttar Pradesh? Obviously, the poor in very large parts of the country feel that the political–official reiteration of concern for them and the goal of social justice no more than lip service.

It is not surprising if social justice is now looked upon by many as pie in the sky in the present economic liberalisation mode of Central Government politics. One may recall that Industrial Policy Statement of July 1991 did not explicitly refer to energy conservation or environmental protection, both particularly important at present stage of development in the country. Energy conservation has been critically important to us for more than 20 years now, because it reduces production costs, and also saves scarce capital (which would otherwise be needed for creating more capacity for energy generation and utilization). As regards natural resources environment as in many developing countries, it provides the bulk of our population its life support. Unfortunately, the impression given by the 1991 Industrial Policy Statement is that these aspects do not loom large in the mind of the government of the day.

Trimming of Government Machinery to Save Resources

V.B. Eswaran[1] has presented an interesting account on the size and functioning of the goverenment machinery presenting the view of the new economic liberalization and structural reforms policies as reinforced by the following speech made by Ex-Union Finance Minister Shri Manmohan Singh of the P.V.Narsingh Rao's Government in June, 1992:

> "Some people have criticized the stabilization programme as being anti poor. I admit that in an economy which has been living beyond its means, stabilization does hurt. It cannot be entirely painless but we have tried to avoid cuts in spending on special sectors to the maximum extent possible. It is true that the fiscal compulsions have forced us to restrain the growth of all expenditure, including social expenditure. But considering that interest payments are a fixed contractual obligation, that defence expenditure cannot be cut beyond a point because of the security environment confronting us, that expenditure on government administration cannot be drastically reduced without a wage and DA freeze or a sharp reduction in employment, that various subsidies cannot be removed overnight, we had very little option but to do what I did. Those who criticize the cuts in social spending should tell us what other expenditure could be cut to make room for increased spending on social sectors."

It is a pity that the Minister was so trite, and soft towards bloated government machinery, for which he did not speak a word. The ever increasing size of the government has now become malignant. With the coalition governments being formed at the Centre and in the states, the expenditure has gone beyond control. The jumbo sized ministries should be the first causality to save resources and revenue. Even as long ago as 1964, a senior secretary to the Central Government had recognized that there were far too many government employees, and halving them would greatly improve the quality and cost effectiveness of the output.[2] Obviously, he had no easy or ready way of translating his perception into practice! Recently, as a consequence of dismantling or reducing controls on industry and business, the public was told that about 2,000 posts in the Central Government machinery would be reduced, but according to the Press, the government has not been able to effect these reductions fully. It must be recognized that the numbers in the government machinery must be reduced and the government must take determined steps for the purpose, such as more efficient working methods, and also 'less government'. This would release resources for better uses. The human aspect of development should *not* mean our government employing unnecessary numbers. The Delhi Electric Supply Undertaking (DESU) has about 27,000 people on its payroll, whereas the Ahmedabad Electricity Company (in the private sector), which is about 40 percent of DESU, in terms of the number of consumers and the load on its system, employs less than 5,000 (and keeps its consumers much better satisfied).[3] This instance perhaps epitomises the wrong conception of government and public sector bodies as mere havens for providing jobs. A glaring example of the district is a fertilizer factory which has remained locked out and non-functional for the last two decades, still has 2,000 odd employees on its payroll.

Cutting Programme Costs

Another aspect of the resource question lies in the assumption in government circles that the cost of a programme or a work is what the departments estimate in their conventional ways of planning and execution. In the rural areas in particular, this need not be the correct or the most efficient method for achieving economy and avoiding wastes. If a job is entrusted to the local community and transparency is ensured, it is generally the experience that the job is done at a lower cost than government estimates, and is of better quality. It is also possible to reduce construction costs by adopting local traditional designs, materials and methods, as is being demonstrated by HUDCO (Housing and Urban Development Corporation), for instance, in recent years. There have been such innovative ventures in the past even by one or two state governments, which happened to be in difficult financial circumstances at the time, and they received public support. The conventional government works departments seem to have little interest in such matters.

The point about mentioning these examples is that one should not, whether inside the government machinery or in the Press, take for granted the present ways of governments. I do not think that one need to calculate, here and now, what resources could be saved out of government salaries and programmes/works to be able to put into practice what has been suggested above. One does need to stress, though, that governments must be persuaded by whatever means to rethink on these matters, instead of merely singing the same song of sacrosanct burdens on the budget, such as interest payments, defence expenditure and the like.

ECOLOGY AND NEEDS OF LOCAL POOR

I would like to quote Manmohan Singh again, the present Prime Minister, Ex-Union Finance Minister, Ex-Governor of the Reserve Bank of India and a noted economist, this time to illustrate the extent of his concern, or perhaps his lack of it, in the medium term, for the natural environment which, I repeat, gives sustenance to most of our people in the countryside. After saying that there is no basis for concluding that the removal of industrial licensing will lead to unchecked growth of 'dirty' and environment-unfriendly industries, he said;

> ".....having abolished industrial licensing, it will be tragic if a new central bureaucracy is created to give environment clearance on a case by case basis. This will introduce excessive uncertainty and arbitrariness in project approval and might negate the beneficial impact of removal of industrial licensing. What is required is laying down of objective standards industry-wise for safeguarding the environment, asking industry to certify compliance with these standards, institution of an effective system of verification and industry audit and heavy penalties for non-compliance with approved environmental standards and norms. Moreover, preference to discretionary controls, wherever possible, the principle of 'the polluter ought to pay' should be made applicable

so that environment accounting becomes an integral part of the project planning."

The Minister seems to refer only to pollution caused by industry and its control. There is more to the matter. For instance, the use of land resources, especially common property, for example, eucalyptus for industrial raw material and for other markets, will deny its use for growing local biomass needs. The commercial or market oriented use of land may earn more cash, but takes away from it the role of sustaining the lives of the local poor. This was the major point of debate in the 1980s whether forests should be assigned the main function of producing raw material for industry–a debate, which the industry lobby lost in 1988 (but revives whenever it senses some sympathy in the Establishment, as in 1991 and again in 1993). The human needs are anti-poor in our country today, as they can go nowhere else for their biomass needs.

Preference to Needs of Local Poor

Another non-pollution aspect of the same theme is that of ensuring an appropriate biomass balance in a 'catchment' from which an industry draws its raw materials, so that local needs are met first. These could be agricultural residues used as fuel and fodder, raw materials for local craftsmen and artisans, medicinal vegetation, small timber, etc.

That the economist's remarks quoted above do not even recognize these dimensions, is a significant pointer to the little relevance the rural poor have his view of the liberalised industrial policy.

Let me quote again from the same speech of the Ex-Finance Minister,

"The overall impact of the stabilization-cum-structural reforms will be to enhance, in the medium term, the ability of our fiscal system to devote larger resources for programme of poverty eradication, social development and environmental protection. This outcome cannot be ensured immediately, but I am convinced that it is only on the basis of a sound fiscal system and a viable balance of payments that one can envisage progress on the poverty front or for the control of environmental degradation..... More emphasis should be laid on provision of services to the poor than subsidies. Similarly, excessive bureaucratisation should give way to greater encouragement being provided for decentralized, location-specific and people-centered initiatives. As presently designed, many of these programmes promote dependence and subservience rather than self-reliance. The acute scarcity of resources provides an opportunity to think in terms of improvement in macro-economic balance, we shall be able to increase expenditure on social sectors in the years."

Thus, not only will the flow of public funds for the poor be contained for an unspecified period but all stops will be pulled to increase the flow of funds and other official help to industry and business. As the urban part of the urban-rural divide, these already transfer resources from the rural area. The new policies also envisage a reduction or abolition of the flow of credit from the banks for rural purposes at concessional rates of interest. Where the rural poor would now obtain replacements of these resources at affordable cost for their betterment, is not clear. It is also not clear whether the Ex-Minister's 'medium term' will be of two or three years, or longer, if expectations of achieving macro-economic balance recede.

These trends of indifference to the poorer sections have been in evidence in some other fields also. For instance, from about 1989-90 the supplies of petroleum products, specially, kerosene have been stagnant in per capita terms in the eastern region where rural population and proportion of the poor are the highest, and the substitute for kerosene has to be obtained by denuding the greenery in the surroundings. As a result, much of the greenery and forest areas in and around Gorakhpur district have been deforested.

Imparting Skills for New Livelihoods

A larger issue is that of alternative livelihoods for rural people, and training and building up skills. In very large parts of the country, the educational system, including the industrial training institutes, has not prepared local young people for moving away from the land-based conventional occupations. What one still sees, therefore, is the usual extensive headloading of fuel wood to markets in Pharenda (A tehsil of Distt. Maharajganj) and other areas surrounding the Gorakhpur District, which are becoming ecologically fragile; this kind of denudation of the tree wealth is intense in times of distress like droughts. The inference to be drawn here is not only that past public investments in these rural areas have not been enough, but that they have also been probably directed into inappropriate activities, quite apart from the question of how efficiently the resources have been used. We may note that these are the poor people, who have depended only on the government social services which have now been short-changed.

Sharing Forest Management with Locals

These days, when the environment and its conservation are much in the air, and there are many non-government organizations (NGOs), some spanning many countries, and at least till now deriving inspiration from the developed world, the question is posed of human rights issues versus environmental protection. Within the government field, this question looms large in the areas of social forests and reserved forest areas such as Campiarganj, Kusumhi Jungle, Pepeeganj etc., which are the part of Gorakhpur District. These areas under forest coverage now occupy over four percent of the area of this region. Officially notified forests cover about 22 percent of the country where the Forest Department is the 'ruler'. The question should be, in our conditions of population pressure on the land and the dependence of the bulk of them on the land, how to ensure that man and his development receive the same as the forests and the denizens of these forest areas. At least in respect of forest management, sustained efforts of informed and determined NGOs have borne fruit to get the governments in India to adopt policies for jointly managing forests with local communities. This was because enlightened officials of the Uttar Pradesh Forest Department proved the value of this approach, and so did the departments of a few other states. For such areas, management of land with a similar approach would be an act of benevolence to the people living in these areas because the laws for these forest areas are harsher towards the local people.

Nothing is more disastrous than the growing perception of the environmental debate being a mere matter of transfer of some technological gadgets and wherewithal to the under-developed and poverty stricken areas of eastern Uttar Pradesh, which has already been a victim of the ingrained step-motherly treatment of the State Government and beauraucracy's economic pattern of thinking and development, based on fossil fuel-intensive technology and industrialization. This metaphysics of economic thinking is based on a view of human nature, which Aristotle some 23 centuries ago and Hobbes some five centuries ago had portrayed as a bundle of never-ending desires and appetites to be fulfilled in materialistic appeasements. As a result, the exorbitant lifestyle and consumption of one-fourth population of the country, at the expense of the poverty of the three-fourths of its population, has been attempted to be maintained by reinforcing and institutionalising existing financial structures on the indices of the dominant ethnocentrism of western capitalist thinking and development. The core of the problem is how to bring about an ethical resurgence in the very economic thinking that has created an antithesis environment and development and put humanity on earth on the brink of extinction.

NEED FOR CHANGE IN BASIC VALUES

From the above analysis, it can be said in unequivocal terms that what is at stake is not the issue of transfer of technologies, economic aid and funds, but the basic concerns which challenge the metaphysics of Union and State Governments' economic and political thinking which has not only damaged the environment but also subjected a vast populace of the country to various economic and ecological hazards.

Man's rights on the environment are, and must be treated as at least the same, as those of other living beings and organisms. Certainly he has duties towards protections of the environment. But his basic needs must first be met. These must be met, as is only just. Not as charity, society generally and the governments particularly, must give the poor amongst us the chance to live with dignity as citizens, not subjects; give them access to good health, meaningful education, peace, and good order; the skills and assets to earn their livelihoods on the land or elsewhere. Neither we cannot see that these are round the corner, nor that our political and administrative institutions now concentrating on these aims.

Economic and Institutional Reforms

A second scenario considered here looks at the possible effects of price increase and institutional reforms. At present, underpricing of electricity is the rule, not the exception in most of Indian states and Uttar Pradesh is not an exception to this phenomena. Such low prices do not reflect improvements in the efficiency with which electric utilities supply their customers. On the contrary, losses during transmission and distribution, partly through theft, run at high levels: 47 percent of generation in India, 58 per cent in Uttar Pradesh and 22 percent in Gorakhpur and other surrounding districts. (In the United States only 8 percent of electricity is lost during transmission, in Japan, 7 percent). These losses, the equivalent of about 75,000 megawatts of capacity and 300 terawatt of hours i.e. 9300 billion kilowatt hours a year, represent a huge loss, which a country like ours can hardly afford. The entire Gorakhpur region barely receives 3 to 4 hours of electricity supply per day throughout the year, thus, badly affecting land irrigation, crop cultivation and other commercial activities too.

The reasons for persistent underpricing are largely institutional. Governments frequently intervene in the day-to-day operations of utilities, and they worry that price increase will exacerbate inflation. Utility managers and their boards may have little say in pricing or investment decisions. Lack of accountability and transparency lead to poor management of the utilities that frequently supply them which is a bigger cause for non-realisation of effective and efficient services.

Because of the need to service the borrowing incurred to build new generating capacity, some states are now starting to raise electricity tariffs. Some are considering or (in a few cases) implementing privatization programmes, usually in the hope of tapping capital markets to build new capacity. Price increase may be easier in countries in which parts of the energy industry are privatized, and management is also likely to be improved. Recently, in Uttar Pradesh a Power Corporation has been established and Uttar Pradesh State Electricity Board (UPSEB) has been trifurcated in order to make power situation efficient and effective. A price hike in the electricity tariff has also been implemented. But, due to rampant corruption and connivance of electricity officials with beauraucrats and businessmen, only the honest and genuine consumers have been affected by the increase in electricity tariff.

Sensible energy prices affect not just the generation of energy but also its use by industry and households. They create incentive for industry to use waste feat for example, through co-generation, which combines power generation with the use of residual thermal energy for other purpose and to improve efficiency in heating, motive power, refrigeration, and lighting. But, these are not implemented with sincerity and honest motivation.

Solar energy can be an effective method of power generation, especially in and around areas of Gorakhpur district. It is available in abundance throughout the year except a few days during monsoons. It is not only helpful in producing cheap and economic power, but also is eco-friendly. Solar energy drives the heat engine that sets the hydrologic cycle in motion as moisture is evaporated and carried aloft. As this happens, atmospheric moisture condenses and falls back to the surface. As a result, some of this water flowing in rivers can be used to generate hydroelectric power, a non-polluting, renewable, and relatively straightforward energy source. Today hydroelectric power, or simply hydropower, produces nearly one-fifth of the world's electricity and about 6 percent of the total energy supply. This model of electricity generation can be potentially utilized in the whole of Uttar Pradesh including Gorakhpur because of the fact that we have perennial rivers having volumes of water to produce power.

Hydropower provides electricity without air pollution of fossil fuels or the sophisticated technology, high costs, and potential radiation hazards of nuclear power. Furthermore, efficient hydropower facilities can be built at the scale appropriate to the site and electric power needs. Finally, since hydropower is based on flowing water, it is renewable as long as there is adequate water to spin the turbines. However, hydroelectric dams located on sediment-laden rivers may have their useful lifetimes shortened as their reservoirs fill with sediment, reducing the volume of water needed to generate power. Even so, most hydropower facilities have expected lifetimes that are much longer than thermal electric fossil fuel or nuclear plants.

TRANSFERRING TECHNOLOGY FOR ENVIRONEMNTALLY-SUSTAINABLE DEVELOPMENT

Industrial technologies and processes that use energy efficiently and work more "clearly" had been a major topic of discussion when world leaders had met in Brazil in June 1992, for the United Nations Conference on Environment and Development (UNCED), popularly known as the Earth Summit taken place at Rio de Janeiro. The focus of the discussion was, to make available technologies to developing countries, at a price they can afford, to enable them to industrialize in a less environmentally-destructive manner than has been done until now in developed countries. Therefore, participatory approach in transfer of technology and economically sustainable development is more important rather than creating big structures or inaugurating mega projects, which not only consume much capital but also threaten fragile ecology.

NOTE

1. Eswaran V.B., *Human Aspects of Developments-Some Reflections, The Indian Jouranl of Public Administration*, July-September, 1993, *Vol. XXXIX, No. 3.*
2. *Ibid.*
3. *Ibid.*

6
Population, Environment and Development

It begs no description that there have been cataclysmic changes at an alarming rate. Deserts are expanding and croplands are losing topsoil due to developmental activities and ever increasing population. Every year some six million hectare of land gets severely degraded and becomes wasteland. The stratospheric ozone layer has been adversely affected with the accumulation of Green House gases and deposition of CO_2 emissions. The result is that the developing countries' natural resource base continues to deteriorate throwing them into deep poverty and uncontrolled pollution, making it highly impossible for them to purchase processed goods and technology or invest in environmental protection.[1]

As studies show, there has been unprecedented growth in the global economy during 1980s. It is much more than that during the last several thousand years from the beginning of civilization until 1950.[2] According to a British environmentalist, with world's population increasing more than three times since 1990, the Gross World Product has increased 21 times, the consumption of fossil fuels 30 times, industrial production 50 times and the average income of the richest one billion people is 20 times -larger than that of the poorest one billion.[3] It seems that the enormous increase in the wealth of the world has not resulted in the equitable distribution of benefits, rather it has widened the gap between the developed and the under-developed.

A study shows that one billion people live in unprecedented luxury and another one billion in destitution. Even American children have more pocket money-$230 a year than the half-a-billion poorest people alive.[4] The environmental deterioration and decline of the economy have fed on one another through the operation of the world market system which has exacerbated this trend and consistently depressed prices of the commodities produced and exported by the developing countries, and aggravated the debt crisis faced by them. Ironically, the fate of the poor and the fate of the planet have become inextricably intertwined. As economist Herman Daly puts it, "there is something fundamentally wrong in treating the earth as if it were a business in liquidation".[5]

The international financial institutions, such as IMF, and the World Bank have been so structured as to perpetuate the colonial economic interest of the rich nations. IMF's insistence on an export-oriented economy, devaluation to boost exports, cut in deficit budgeting and severe austerity measures have impelled the Third World, in their bid to earn more foreign exchange, to exploit their environment and adopt economic policies with only short-term gains in mind. The World Bank continues its traditional lending policies to the Third World for large capital-intensive projects without taking into account the rural poor's resentment and opposition. This is despite the recent opening of a new environmental department in the World Bank.

There is a growing tendency to define development just as amassing wealth or the conquest of extraordinary technological power and skill which is the sole cause of persistent poverty and of human rapacity to spoil nature. Philosopher Hans Jonas, in his *The Imperative Responsibility*[6] puts forward arguments to combat this very technological and economic thinking, for an ethics of responsibility for the cosmos. The crux of the debate is that the ethnocentrism of the dominant western economic thinking and pattern of development does not take into consideration the natural threshold, the carrying capacity of the Planet Earth and ecological costs of production. It has resulted in a Western masculine metaphysical understanding of power in terms of dominance over nature, control over economic activities and territorial states. Can such an economic development be called successful, if it has affected a terrible despoliation of nature and environment and destitution of the people in the developing country like us? This is what led many to denounce it as antithesis of authentic development or antidevelopment.[7] Recent years have seen the Third World obsessed with ever enlarging economic growth.

Development Not by GNP Alone

Many world organizations such as the World Bank, OECD and UNTACD have been ranking nations on the indices of aggregate economic growth measured in terms of GNP. The authors of the first annual report of the UNDP on Human Development have challenged this in 1990, addressing the need to transcend the narrow economic vision of development.[8] Blind faith in the free market determined price mechanism to measure economic costs of resources is highly imperfect. The predominant stress of most neo-classical economists is only on goods or services that are either inputs or products of economic activities. This is what has led one authority to opine: "What is the cost of a tree which cleans air, regulates water flows, fixes nitrogen, absorbs pollutants, produces biomass and provides habitat to other flora and fauna? At best, economists can complete 'replacement costs' of those functions that are replaceable....... But what happens to the irreplaceable?"[9]

Need for New Indices of Development

The desideratum is to replace growth obsession economics. There has already been a burgeoning resurgence of new indicators of progress, demonstrating mounting disenchantment with the customary habit of equating progress with economic growth. Herman Daly, a World Bank economist, and theologian John Lobb of the Claremont School of Theology, have developed an "Index of Sustainable Economic Welfare" (ISEW). The UN has devised the Human Development Index

(HDI). The third index is the Grain Consumption Person, a very sensitive measure of changes in well being, in low-income countries.[10]

Resurgence of ethical values and restoring them to a central place in the discipline of economics has already been noticed in the works of eminent writers. Amartya Sen's *On Ethics and Economics*[11] and Herman Daly's *For The Common Good* are noteworthy. Sociologist, Amitai Etzioni, in his *The Moral Dimension*[12] is of the view that apart from rationality, which should not be confined to a mere benefit-cost calculus of maximum self-interest; altruism and moral duty are equal motivators of human behaviour in the economic realm. Paul Ekins and his colleagues have launched 'The Living Economy Movement' which stresses the exploration of economic requirements of the common good and community building.

The ethical revolution in economics is still in a nascent stage. Unless and until the prevalent trends and tendencies of current economic development are reversed on suggested ethical lines, the continued antithesis between development and environment cannot be replaced by a new ethics of sustainable development. In the brilliant words of an authority: "It may well be that the kind of development needed is one that is open to transcendence in order to relative economic growth and technological progress as values, as well as to de-absolutise the claims made by managerial absolutism and the modern nation-state".[13]

In the economic substructure, what is needed is an ethical resurgence that would release policy planners and economists from a growth obsession, finding essentially a dichotomy between goods and good. It insists on developing alternative measures of progress based on equity and sustainability. What an authority addressed to Latin Americans some years ago deserves to be quoted :

> "Does man lives by GNP alone? Should we not take advantage of our long-term vision and ask what kind of person Latin American may wish to evolve by the end of this century? What are the transcendental values cultural, ethical, artistic, religious, morals

that extend beyond the current working of the purely economic and social system? How to appeal the youth, who so often seek nourishment in dreams, as well as in bread?"[14]

John Maynard Keynes, father of neo-classical economics, and the chief architect of the post-World War II financial institutions designed at Bretton Woods, proposed a toast "to economics and to economists who are not the trustees of civilization, but of the possibility of civilization".[15] That means Keynes attributes a purely instrumental value to economics, which would in contemporary idioms mean creation of the possibilities of human development. Adam Smith, the father of classical economics in his *The Theory of Moral Sentiments*, assumed that a higher providence ordains that the working of economic laws would not exceed certain moral bounds and ethical precincts and result in total selfishness.[16] To show a beacon of light to man who is already in an environmental *cul de sac*, the need is not to trivialize the basic issues and concerns by putting emphasis on mere transfer of technology or funds. It necessitates, as discussed above, a rethinking of and a revolution in dominant Western ethnocentric paradigms of development.

Programme for Poverty Alleviation and Environmental Regeneration

It is now widely accepted that mass poverty is a major factor in environmental stress and degradation in the under-developed regions like that of district Gorakhpur, even though many see poverty in isolation and overlook the political, social and economic causes, national and international, that have given rise to it.

Poverty, population growth, and lack of employment opportunities feed upon each other to increase pressures on the environment, while the shortage or lack of resources precludes any long-term investments being made to improve the productivity of natural resources. Paralleled with the quest for solutions to the underlying causes of poverty, it is still possible to undertake an immediate and focussed regional assault on some of the worst aspects of local poverty and environmental degradation.

In many of the world's degraded ecological areas, experience has shown that environmental regeneration activities such as the control of soil erosion and desertification, afforestation and grasslands development, and soil and water conservation can greatly improve land productivity and village ecosystems. As these activities are highly labour-intensive, they have the potential of providing opportunities for the poor to earn cash incomes with which to fulfil some of their basic subsistence needs, particularly if undertaken with adequate financial resources and carried out under the control and responsibility of local communities. It is thus possible to devise poverty-alleviating, employment, rehabilitation and development programmes that can ensure that the labour of millions of people is harnessed to regenerate the environment and secure a better future for the local populations.

Such programmes can also play an important role in dealing with major global environmental problems such as global warming and biodiversity conservation. A global programme for poverty alleviation would make it possible to "green" large areas and thus help to absorb substantial amounts of the carbon now accumulated in the atmosphere. And by providing employment opportunities in activities that aim to bring about ecological regeneration, it could substantially reduce the existing human pressures on areas that are rich in biological resources.

Some estimates show that such a programme would not cost more than US $40 million per year, a relatively modest sum compared to the global sums spent on luxury items or on armaments. Innovative international schemes for financing the global programme of poverty alleviation and environmental regeneration would need to be worked out, including tapping of the resources released through disarmament and a reduction of military expenditure.

Institutional arrangements to implement, and to finance and monitor the operation of, agreements reached at the Rio conference were among the key subjects of negotiations in UNCED. Two principles should underline the formation of new institutional arrangements:

1. Democratic governance and transparency within the UN system, and
2. An integrated vision of a problem that has hitherto been approached in a piecemeal fashion.

Moreover, in view of the vast resources needed for carrying out a truly global environment-development policy, care should be taken on the one hand, to ensure that the efforts do not compete with the traditional call on the World Bank's resources and, on the other hand, to ensure that novel means and modes are devised for fund-raising and financing such a policy including those which would not cause the developing and underdeveloped countries to incur further heavy debts.

Population Pressures

In any global equation, human beings are "wild card", it is said. As primary consumers of the earth's natural resources, people are both agents of development and victims of environmental degradation. They are also, potentially, the means of conservation. Any development has an impact on the ecosystem. It would be naïve to think otherwise.

"The relationship between population dynamics and the ecosystems on which the survival and well-being of people depend will be decisive in achieving sustainable development", says Maurice Strong, Secretary-General of the United Nations Conference on Environment and Development (UNCED), held in Brazil in June 1992.

World leaders convened at Rio de Janeiro for the Earth Summit had been asked to consider the complex links between demographic trends, environment and development. Specifically, they were urged to draft measures to stabilize global population and to support each country in achieving a sustainable balance between its populations, its resources and the quality of life to which its citizens aspire.

For policy makers, the issue is how demographic pressures relate to economic growth, consumption patterns, poverty and environmental quality. With economic growth comes increased consumption. As developing countries produce and consume more of the basic necessities than the developed world, now takes for granted from adequate food and housing to refrigerators and cars the challenge will be to make such consumption sustainable. Growing human numbers pace greater demands on land and water, generate waste and pollution, and encroach on the habitats of non-human species that share the biosphere. Population interacts with every major issue affecting the health of economies, people and the planet from faltering gross domestic product (GDP) to forest loss and global climate change.

Actually, the population issue is a bit more complex than this. Even with merely 'possible'persons out of the picture, population policy can involve a conflict between generations. To some degree, it is in the interests of certain segments of present generations to have unrestrained procreative liberty, whereas it is generally in the interest of future generations to have earlier population growth limited. Apart from the emotional satisfaction produced by children, for example, there are often concrete reasons why parents in agrarian societies opt for numerous offsprings. At the same time, larger family size can disadvantage the children themselves, a fact that has led some demographers to speak of the "parental exploitation of children" in the underdeveloped setting.

The Liberation of the Environment

Environmental sins and sufferings are not new. Humans have always exploited the territories within reach. The question is whether the technology that has extended our reach can nowalso liberate the environment from human impact and perhaps even transform the environment for the better. The answer is that well-established trajectories, raising the efficiency with which people use energy, land, water, and materials, can cut pollution and leave much more soil unturned. What is more, present cultural conditions favour this movement.

Attitudes worldwide toward nature, and perhaps inseparably toward one another as humans, are changing. "Green" is the new Mantra and "Ecology" is going to be the new religion of the humanity. Jungles and forests, commonly domains of danger and deprivity in popular children's stories until a decade or two ago, are now friendly and romantic. The Amazon has been transformed into a magical place, sanctified by the ghost of Chico Mendes, the Brazilian rubber tapper. Environmental shrines, such as the Great Sacrophagus at Chernobyl, begin to fill the landscape. The characterization of animals from wolves to whales has changed. Neither the brothers Grimm nor Jack London, could publish today without an uproar about the inhumanity of their ideas toward nature and I would add, with regard to gender and race as well. Popularity of 'Harry Potter' is a testimony to this fact that nature and its supernatural aspects still attract wide scale attention.

Finally, after a very long preparation, our science and technology are ready also to reconcile our economy and the environment, to effect the Copernican turn. In fact, long before environmental policy became conscious of itself, the system had set de-carbonization in motion. A highly efficient hydrogen economy, landless agriculture, industrial ecosystems in which waste virtually disappears. Over the present century these can enable large, prosperous human populations to co-exist with the whales and the lions and the eagles and all that underline them if we are mentally prepared, which I believe we are.

NOTE

1. Gaan Narotham, Environment and Development : A Search For An Economic Re-thinkin, *The Indian Journal of Public Administration*, July-September, 1993 Vol. XXXIX, No. 3.
2. See, *State of the World 1991 : A World Watch Institute Report on Progress Towards a Sustainable Society*, New York, World Watch Institute,1911, p. 6.
3. Norman Myers, *The Gaia Atlas of Future Worlds*, New York, Anchor Books/Double Day, 1990.
4. OECD, *The World Economy in the 20th Century, Paris*, OECD, 1989;, Alan. B. During "Poverty and the Environment : Reversing the Downward Spiral", *World Watch Paper 92*, Washigton D.C., World Watch Institute, November 1989.

5. Herman E. Daly, "Sustainable Development : From Concept and Theory Towards Operational Principles", *Population and Development Review* (Proceedings of Hoover Institution Conference on Population and Development).

6. Hans Jonas, *The Imperative of Responsibility*, Chicago, the University of Chicago Press, 1984.

7. Dennis Goulet, *The Cruel Choice : a New Concept in the Theory of Development*, New York, Atheneum, 1971, pp.215-35.

8. Mahbub Ul Haq and Paul Streeten, *Human Development Report 1990* UNDP, New York, Oxford University Press, 1990.

9. Shekar Singh, "The Global Environmental Debate", *The Indian Journal of Public Administration*, Vol. XXXV, No. 3, July-September 1989, p. 371.

10. Herman E. Daly and John B. Lobb, Jr. *For the Common Good: Redirecting the Economy Towards Community, The Environment and a Sustainable Future*, Boston, Beacon Press, 1989; and U.N. Development Programme (UNDP), *Human Development Report 1990*, op.cit. Also see Robert Costanza and Lisa Wainger, *"No Accounting for Nature : How Conventional Economics Distorts the Real Value of Things"*, Washington Post, September 2, 1990, p. 83.

11. Amartya Sen, *On Ethics and Economics*, Oxford, Oxford University Press, 1987.

12. Amital Etzoni, *The Moral Dimension*, New York, the Free Press/ Mac Millian, 1988.

13. Denis Goulet, "International Ethics and Human Rights", *Alternatives* Vol. 17, No. 2, Spring 1992, p. 241.

14. David H. Pollock, "A Latin American Strategy to the Year 2000 : Can the Past Serve as a Guide to the Future?' in David H. Pollock and A.R.M Ritter (eds), *Latin American Prospects for the 80s: What Kinds of Development, Reformation of Transformation?*, Ottawa Norman Patterson School of International Affairs, Carleton University Conference Proceedings, November 1980, pp.1-37.

15. Cited in Benjamin Higgins, *Economics Development, Problems, Principles, and Polices;* New York, N.W. Norton, 1968, p. 3.

16. For Details on how Economics Gradually Eliminated Values, see Kenneth Lux, *Adam Smith's Mistake*, Boston, Shambela, 1990.

7

Political Dimensions of Development

It is only a matter of time before the concept of sustainability through misuse becomes meaningless. There was a time when for countries like ours, coming out of a long history of colonization, sustainability had another meaning it was a fundamental socio-political concept. It meant self-sustained growth and self-reliance. In that sense it was a far more powerful concept and a motive force than the concept today that has become dependent on the sculptor.

Almost every major international gathering and a wide variety of new initiatives from the official Rio conference[1] to the setting up of the Business Council for Sustainable Development, from the annual World Bank-IMF meeting to new projects of national governments seemingly accept that environment and development are compatible and that sustainable development is compatible with current patterns of achieving economic growth and political development.[2]

Despite the multiple interpretations associated with the terms, the notion of sustainability applies not only to the replenishable use of renewable resources (a process of natural resources use where the rate of take is equal to the rate of renewal or regeneration) but to a much broader concept, which embraces ethical norms pertaining to the survival of the diversity of living matter, to the rights of future generations, and to the definition of institutions from the local to the global responsible for ensuring that such rights are integrated into policies and actions. In that sense, the concept of sustainable development would only have meaning if it is recast to encompass issues of justice, democracy and ecological sustainability.[3]

In countries of the Third World, community mobilization around predominantly ecological issues is centuries old.[4] For instance, in numerous parts of India, by the beginning of the 19th Century, communities had mobilized to oppose colonial intrusion, which attempted to transform local eco-system. For instance, this opposition was significantly manifest in tribal areas where these communities did not acquiesce quietly in the face of external intervention.[5] There were protests and rebellions against colonial laws, such as the Forest Act of 1876. While the newly emergent colonial science of forestry had begun to blame the agrarian system of forest dwellers for decline in forest coverage, tribal peasants were waging struggles against state intervention in forest resources based on their own moral economy. Guha and Gadgil have aptly described this conflict as between "the political economy of profit and the moral economy of provision".[6] And, in the recent past, with the ecological and social impacts of rampant industrialization manifesting themselves in dramatic tragedies ranging from Bhopal to Amazonia, from the decimation of the forests of Asia to the impacts of nuclear testing in the Pacific, from mini Chernobyl's and Exxon Valdez's to unmanageable piles of pollution and toxic wastes, major questions on the sustainability of the dominant patterns of economic development are being increasingly asked[7].

The Threats : The Changing Configuration of Economic Power

Global capital has been increasingly mobile across national boundaries. The global economy continues to be substantially dominated by the industrial countries with selective and tactical alliances with elites in the industrialized world. This extractive economy is geared primarily for the benefit of elites who control world trade and provide fiscal incentives to invest in export-oriented extraction. In fact, for most parts of the Third World, there has been a declining control that local producers have on what they produce, how they produce and at what price they can sell it in the market. Over the past decade and for countries like India more intensively over the past two or more years, when the process of disinvestments began, this control has even started to slip from the hands of national elites.

It would be naïve to assume that institutions like the World Bank are "developmental institutions". There is now overwhelming evidence that there is an emerging nexus between four factors-Transnational Corporations, Multinational Banks, GATT, and economic elites in the so called North and South to control the global and national economies. In that sense, there have been fundamental changes in the global political economy.

Fifteen of the world's largest Transnational Corporations (TNCs) now have gross incomes that are larger than the GDP of 120 countries. Over the past decade, it has been these TNCs (directly and through collaborations with domestic capital) that have succeeded in exerting enormous power over national governments. They have been the main agencies that have aggressively pushed for deregulation and privatization not only of national economies but also in the world economy as a whole. Implication of this shift in the configuration of power has received little attention from either scholars or activists. The Indian Government's former Chief Economic Advisor stated last year that TNCs are emerging as the new global government, a World Inc. Ltd., with the G-7 as the Board of Directors.[8] This is not to say that these corporations have an uncontentious relationship between each other or that this structure is already in place or that the traditional system of competitions and mutual suspicion has been subsumed in this "global project". Far from it, what is being highlighted, however, are the new arrangements whose skeleton and some elements of the form are already evident.

In countries like India, this control and efforts to direct the national economy have also meant withdrawing protective legislation that protected the domestic industry, the indigenous producer as well as other occupations that were better suited for the specific context, *i.e.* that a society was labour-intensive.

The implications of this go beyond notions of national sovereignty. The emerging thinking argues for a restructuring of the United Nations system itself. Specifically, it is being propagated that the UN should henceforth confine itself to the social and cultural terrain with the economic issues being controlled by an institution like the Multilateral Trading Organization as proposed in the Dunkel Proposals and by a Council representing the nexus.

A careful study of the GATT proposals and in particular, the Dunkel Proposals[9] and the process that led to their drafting demonstrates almost total exclusion of most governments of the Third World, with no participation of independent citizens' groups or other non-governmental agencies. In fact, the only non-governmental actors that were permitted access in most of the deliberations were senior representatives of TNCs.

Undoubtedly, these proposals legitimize a fundamentally undemocratic trading regime that will lead to the acceptance of the lowest common denominator in environmental standards and an inevitable escalation in the violation of human rights, as governments are coerced into accepting the terms of the new global regime. Even legislators in the US are concerned about the implications of this lowering of existing health, safety and environmental standards.[10]

Threats to Federalism and Sovereignty

For countries such as India, these proposals and the emergence of a more coordinated nexus between global economic institutions and national political and economic elites, would pose a direct threat to its federal structure since the nation would be brought under one homogeneous economic regime, whose terms would primarily be set by this quarter of factors. In fact, the threat to sovereignty, inherent in instrument such as GATT, would itself be a significant factor to compel rethinking on GATT.[11]

This critique should not be misunderstood as an argument against globalization *per se*. It is nevertheless the contention here that economic globalization would only help the country as a whole when trade, labour and capital flows *are between equals* and even here there can be tremendous problems as the current differences between the US, Europe, and Japan indicate. It is indeed a pernicious philosophy that with globalization, the market becomes available to all parties. In a differentiated heterogeneous world both within and between countries, globalization can only be equated with the slow subjugation of weaker sections and the annihilation of cultural and ecological diversity.[12]

The current agenda for liberalization in the absence of a fundamental social and economic restructuring to a more democratic and accountable system are then just a means to the ends outlined above in the emerging strategies of the nexus. The imposition of these macro policies is then not just an imposition of monoculture but, a fundamental flattening of the spaces for sustaining the struggles for ecological sustainability and democracy. Clearly, proposals like GATT as well the directions of the new alignment of economic power represent an agenda for the reconciliation of the world.[13]

Internally, within each country, the social and environmental impacts of these processes are far-reaching and grim. Of course, these impacts are differentially felt, both at the level of national economies as well as at the individual level. Yet, since most countries in the Third World are not implementing a democratic agenda that includes significant internal restructuring, liberalization impels a maximization of the export of primary goods and natural resources. Since these patterns are dependent on the extraction of these resources, there is a perpetuation of processes of internal colonization. The resultant escalation of the loss of control over productive resources and the consequent increase in economic and social insecurity contributes to the growth of, and the escalation of social conflict. As in the case seen recently, cane growing farmers of Gorakhpur and other surrounding districts of Uttar Pradesh, set on fire a sugar mill in Munderwa, on 15/16

December 2002, where police had to resort to firing and lathi charge in which many farmers were killed. The whole dispute was related to a political decision, where the Union Government had to decide minimum support price of the sugarcane in Eastern Uttar Pradesh. It is alleged that Union and the State Governments favoured the interest of sugar mill owners. Thus, it was a political decision having economic implications.

Secondly, national governments, particularly in the Third World, have been militarizing themselves not just as a check to threats from other national aggressors but also as a means to contain internal unrest and militancy. Regional liberation movements, collective dissent and other social conflicts endanger greater internal militarisation and counter-insurgency. It is a point to be noted that the boarder areas of eastern Uttar Pradesh and rural mass base of Gorakhpur district have become a centre of Maoist and armed communist activities, in recent 5 or 6 years. To quell this, even a special boarder force of the State Armed Police has been instituted under the supervision of an Additional Director General rank police officer. A huge amount of budgetary allocation has been spent for modernization and arms purchase for this force, which a poor state like us can hardly afford. This, in turn, absorbs key resources and disrupts the possibilities of long-term culturally and socially sensitive sustainable development. This reality is a central part of the "development question".

Other Costs of Economic Development

As any honest review of the present juncture would conclude, the current patterns of achieving economic growth are incompatible with social justice and with sustainable ecological principles.

To reiterate, the dominant patterns of economic development continue to underplay or overlook the following endemic problems.

1. This economics continues to be reductionist. For instance, the justification of a wide range of development projects is still based on a cost-benefit analysis where it

is assumed that everything is quantifiable. While important work is being done to incorporate hitherto non-quantifiable elements (e. g., the value of the forest to a local forest-based community), there continues to be an assumption that it is only a matter of time before sophistication will be achieved to comprehensively account for such costs. Can the true value of an old-grown forest be calculated? Can the loss of culture or community cohesion be calculated? Additionally, the dominant processes of evaluation identify natural resource utilization merely with extraction, thereby ignoring the short and long-term productive function of conserved resources. This reductionism has been the direct consequence of a gradual process over the past century, where there has been a disembodying of economics from its ethical and cultural moorings.[14]

2. The faith in current strategies of achieving growth continues to be premised on the assumption that Nature is a limitless resource to relieve hardship and inconvenience experienced on "Man's estate". Little seems to have changed since the Hobbesian contention (paralleled by others like Bacon) that the "real destiny of natural goods is to be found in their economic utilization". More recently, a historian of science stated that the current strategies of economic development are predominantly based on "the idea that nature and its resources exist solely to be used for the material benefit of humans and for the particular profit of those who own them.".[15]

3. The dominant strategies continue to be dependent on the intensive and extensive exploitation of natural resources. Millions integrally depend on the health of these systems and any disruption is a direct undermining not only of the subsistent economies of these communities but also of their livelihoods, life styles and lives themselves. Much of the diversity and complexity of social and ecological existence on the planet is due to the rich mix of communities, landscapes,

and ways of social organizing that the human species through its evolution has preserved or helped create. This diversity has contributed to the evolution of complex human consciousness a diversity that is being fundamentally undermined by the monoculture of modern industrialism and consumerism.

4. There continues to be apathy and indifference at the rapid conversion of ecological capital into economic capital with the terms of conversion shifting from the local producer to the city trader, from the local producer to the national centre, and from the local producer to those who control global markets and global trade. There is in that sense a progression in the loss of control from the local to the global on terms that the local rarely sets.
5. Development, by and large, continues to identify with sectoral growth, ignoring underdevelopment induced in related sectors through negative externalities and related undermining of the productivity of the ecosystem.
6. Dominant economic policies continue to identify economic value merely with exchange value of marketable resources, ignoring the use value of more vital resources and ecological processes. (A resource may have a very high use value but little or no exchange value.) As a corollary, natural resources can have economic utility that cannot be quantified through the exchange value in the market.
7. There continues the perpetuation of one ideology of modernization: a supposedly shared public ideal of increasing material abundance and consumption through increased technological and social control by corporations and the state. This ideology rationalises both the unequal distribution of wealth and power and the destruction of the environment.

Most development planning continues to be based on these realities and thus neglects eco-systemic linkages as well as natural processes, which are vital and critical for the subsistence and survival of millions of people on the planet.

Whose Common Future

The critical interrelationship between environment and development impelled the UN to set up the Commission on Environment and Development (headed by Gro Harlem Bruntdland), to "propose long-term environmental strategies for achieving sustainable development by the year 2000 and beyond". Its report, *Our Common Future* became one of the most influential documents in the sustainable development debate. The report exhaustively documented how current strategies of achieving economic growth were dependent on the increasing use of natural resources, a vast number of which in turn were either reaching levels of critical shortage or were being degraded to the point where they jeopardized the very growth that they had generated. Also documented was the wasteful over-consumption of industrialized countries and of elites in the less industrialized ones. Arms expenditures and overpopulation were other causal factors contributing to the downward spiral of degradation and poverty. Ironically, after such graphic accounting of the adverse impacts of rampant industrialization, affluent lifestyles and militarisation, the recommendations paralleled conventional developmental thinking. "If large parts of the developing world are to avert economic, social and environmental catastrophes, it is essential that global economic growth be revitalilsed".[16]

The Commission's conclusion fundamentally contradicted its own exhaustive data as well as its analysis that over-consumption and growth are the root cause of the problem. Political and economic expediency seemed to have overridden the clear implications of its own analysis. There was in that sense a convergence with the unambiguous faith of institutions, like the World Bank which continue to believe and propagate that increasing economic growth and integration of the world economy are the determinants of development and progress.[17] In fact, the Bank's belief is that growth achieved through a primarily export-oriented industrial economy is a precondition for national governments releasing surplus funds for "environmental protection".

On the other hand, the UNDP's 1992 and 1993 Reports [18] differ in emphasis. They stress the need for more equitable distribution of "economic assets and opportunities" as well as other human indicators than GNP and GDP to judge the health of societies. However, the fundamental issues of global inequity, of the critical links between economic growth strategies and environmental degradation, the downward spiral of environmental degradation and its umbilical link with the creation and compounding of poverty were neglected.

Another challenging critique of the dominant world-view comes from a World Bank economist, Herman Daly. In his *Economics, Ecology and Ethics* and in *For The Common Good*, he presents an eloquent argument of how to use natural means rationally in service of moral ends—which is stated in this article as being an unequivocal prerequisite in the struggle for equity and democracy. Daly locates one of the roots of the present malaise in Keynesian growth economics, which justifies the need to maximize material and energy-resource utilization without attention to the constraints. His argument for a steady-state economy is not a defence of a no-growth economy but a policy of selective growth governed by ethical consideration, for example, augmented economic growth for the poor and a diminished growth for the rich.

Ecological Degradation and Poverty

For countries like India, the dominant model of economic development, which continues to define the utilization of natural resources, evolved in a colonial context when massive economic growth in the Western world was generated at the expense of creating underdevelopment in the colonies. This is now a well-known, unacknowledged fact. When countries that were colonized became independent, they adopted the same pattern of resource extraction and use generating in the process development and under-development within each country. For most of the Third World, it was not feasible to colonise other parts of the world to promote growth. Inevitably, therefore *they had to colonise internally* to sustain growth in the industrial and urban sectors.

Ghandhiji had predicted the dangers of doing this when he had said, "God forbid that India should ever take to industrialization after the manner of the West. The economic imperialism of a single island kingdom is today keeping the world in chains. If, an entire nation of 300 million took to similar economic exploitation it would strip the world bare like locusts".[19]

Uttar Pradesh in India, especially Gorakhpur district, is a classic case of uneven development. Part of this was encoded in the post-independence emphasis on rapid industrialization and part of it was a direct consequence of not seriously addressing the endemic issues of social inequality. A Western modelled industrialism has not only created and reinforced underdevelopment but also led to serious erosion of eco-systems, which were the fundamental basis of a majority of people. "The poor live within a biomass-based subsistent economy..... All their fundamental needs (food, fodder, fuel–firewood, cowdung, crop-waste, fertilizers-dung manure, leaves litter, building materials, timber, thatch, and herbal medicines) are collected (often freely, from the immediate environment). To these be added water, which though not biomass itself is biomass related and crucial for survival".[20]

The Gorakhpur case forcefully illustrates the integral link between land degradation and poverty. In fact, evidence from countries around the world[21] suggests a link between processes of degradation, erosion of social and cultural fabric of local society and consequent escalation of social and economic insecurity. Further, this process contributes to a variety of social conflicts.

The integral link between ecological degradation and poverty and conflict can be gleaned from the following facts, as derived from research and survey of Gorakhpur district.

1. Out of 3483.8 sq.km. of land mass, about 586.5 sq.km., is now considered degraded in one form or another.

Annually, due to unsustainable practices, approximately 50 sq.km. is going out of production. Much of this is due to deforestation, where approximately 8.5 sq.km. of forests are cut every year, primarily for industrial and commercial purposes.[22] Of course, there are other contributing factors-bad land and water management, submergence in dam reservoirs, industrial and urban expansion, mining, overgrazing of declining grazing lands, wind and water erosion (compounded by deforestation), flooding and water logging. These processes contribute to a loss of productivity in another 10 sq.km. land annually. Wind and rainwater are the main agents of soil erosion accentuated by deforestation and excessive grazing. The problem assumes alarming dimensions in the catchment areas of the rivers, rivulets and nullahs, which are quite abundant in the district and where more and more land is fastly and steadily being devoured by the erosion and changed into ghastly barren wastes. It was estimated that a total area of nearly 52,500 hectares of land is affected or threatened by soil erosion in the district in some form or the other. The following statement gives the area of land affected by soil erosion in the seven tahsils of the district.

Table 7.1

Tahsil	*Area affected (in Hectares)*
Gorakhpur Sadar	2,500
Basgoan & Gola	20,000
Sahjanwa & Khajani	12,000
Chauri-Chaura & Kampierganj	18,000
Total	52,5,00

To prevent soil erosion a unit of soil conservation department was established in the district in 1968. Since then the work is going on in the catchment areas of Turran Nala, Pharen Nala, Majhna Nala, Rohini River, Ami River and Powa Nala. Besides, the work of levelling, construction of dam-cum-check roads, check dams, contour dams is also being undertaken as an integrated approach towards conservation of soil. Methods mostly applied to combat soil erosion in the district are contour bunding, levelling and check-daming. Further, this contributes to desertification and creation of drought-prone conditions. In years of inadequate and irregular rainfall, the intensity of drought is more severe. These trends of how human consumption and life style contribute to drought and famine, though have not been properly documented in the district Gorakhpur, yet a few statistical and relevant pieces of information can be examined.[23] In areas like the western Gorakhpur, the processes of drought and desertification have set into motion an ever-expanding cycle of degradation, to the point where recovery and regeneration of land are either impossible or an enormously expensive enterprise.

2. Further, conflicting administrative approaches and gross inequality in landholding inevitably worsen the plight of those who are dependent on the land for their daily subsistence.[24] According to the Ministry of Agriculture, this process of degradation amounts to a loss of between 3 and 5 million tonnes of foodgrains only in the Eastern Uttar Pradesh especially in and around the district of Gorakhpur.
3. Given these processes and the resultant decline in livelihood for the millions critically dependent on these resources, there are few options but to cultivate increasingly marginal lands thereby compounding the sustainability crisis. The impact on women is even more severe, as the loss of access to fuel, fodder and water forces them to walk miles to collect the essential necessities for their subsistence.[25]
4. The consequent escalation in the pressure on available arable land is so enormous that it has contributed to the growth of criminalization in the countryside with illegal

occupation of community lands, the formation of "land armies" by landlords to oppose any demands by marginal farmers for land, as well as the increasingly militant movements to assert local control over productive resources like land and forests. Since the decade of 1980s there has been an increasing tendency in land grabbing cases as well as unauthorized possession and ownership of the urban as well as rural land, which has also increased in the last 20 years in the district Gorakhpur on behest and insistence of political leaders, mafias and other anti-social elements in collusion with the district administration. The latter process, most often results in state repression compounding the climate of social unrest. In the year 2000, a Cabinet Minister in the State Government is said to have grabbed more than two acers of prime land in the heart of Gorakhpur city, in connivance with the district administration. After completion of the inquiry, the state High Court and the 'Lok Ayukta' have even passed stricture against the Minister and the erring district magistrate.

5. Much of this has severely strained social relations within communities and between communities and the state. For instance, every year, over 5,000 people are forcibly dislocated by developmental projects in the eastern Uttar Pradesh.[26] Most are not rehabilitated, and alternatives are rarely provided. In the process, communities and families are broken up, destroying structures of social and economic support. The simultaneous loss of cultural diversity is an inevitable consequence. For instance, a report based on a comprehensive survey of people displaced by the construction of dams, development of Ramgarh Tal Project and laying down of new residential colonies in the outer areas of Gorakhpur city states: "Many of the oustees, particularly the rural people, have fallen into the typical cycle of debt bondage, coupled with increasing destitution and intermittent employment as contract labourers in coal mines and

elsewhere... most were simply kicked out with nothing left to fend for themselves".[27]

Since 1947, one of the central policy directives of the government has been its concern with poverty alleviation. The economy has also grown steadily. From 1960-1990, the Net National Product (NNP) increased from 1.3 to 3.7 per cent while the per capita NNP increased from 1.4 to 2.0 per cent. Yet, the per centage of people below the poverty line has remained constant, with "40 per cent of the population living below poverty line and another 20-30 per cent just above it".[28] Despite a majority of years where growing rates of over 3.0 per cent were achieved, and despite a steadily growing GNP, the share of the bottom 10 per cent of families plummeted from 2.5 per cent in 1961 to 0.07 per cent in 1991, a drop of 30 times. Stated differently, in this 30-year period, the incomes of the richest 20 per cent doubled from 30 times to 60 times of the poorest 20 per cent. It is clear that as many have remarked and as is documented quite powerfully in the past two reports of the UNDP, one of the major costs of the past decade of economic growth is greater inequity.

Under the new regime of economic liberalization, India lost 6.6 million jobs in the years ending March 31, 1993. Since the formal adoption of the Structural Adjustment Programmes (December 1990), 23 million jobs have been lost.[29] Almost across the board, there have been cutbacks in social service spending ranging from 20 to 80 per cent. Several crucial programmes have been withdrawn all together.[30]

The poor in most of the Third World (a majority of whom, remain rural) can only be raised above a life of poverty by recovering and regenerating degraded lands and by equitably distributing these resources. However, such developmental strategies are almost totally absent, in today's official developmental plans.

Challenges of Some Contemporary Movements

Over the past two decades, in India as in countries across the planet, there has been a virtual explosion of movements, group and individuals who are attempting to redefine democracy and justice from a self-consciously ecological perspective.[31] Embracing a staggering variety of alternatives, these cultures of resistance are posing a fundamental challenge not only to the dominant patterns of economic development, but also to the conventional notions of environment, development and equity.[32]

The sharpening of resistance has come in the wake of the preceding reality and a realization that if the issues of equity, democracy and sustainability have to be seriously raised, they have to confront hegemonic economic and political power.[33] This is not the place to present the diversity of social movements who are struggling to define an ecological politics. [34]

Many increasingly feel that the state has reneged on its constitutional responsibility. Legislation and an externally conceived administrative system have legitimized the disruptive forces. For instance, laws like the Forest Act effectively make forest dwellers criminals. In the name of forest management, successive governments with negligible change in the post-Independence period have contributed to the erosion of the relatively symbiotic human-nature relationship in most rural areas and legitimized the rapid conversion of nature into commodities. To the state and to the external capitalist interests, the forest was only as valuable as the economic value of commercial timber and forest 'products'. In that sense, the forest became one more factory. Tradition and custom, the older system of regulation and use were slowly rendered meaningless and in some cases consciously destroyed.[35]

Throughout this period, these processes have not been quietly accepted by the communities. There were sustained protests and rebellions against the politics of the colonizer and the intrusions of the 'outsider'.[36] Most of them were brutally crushed by the state. Today's governments, by and large, continue to be part of this tradition. What, over the past two centuries, have been issues highlighted by these movements? Some of them can be enumerated as such,

1. Almost without exception, every rural community has experienced a decline in the access and control over productive natural resources-land, forests and water. With rare exceptions, all customary rights were thus de-legitimized. Modern industrial development with its insatiable appetite for natural resources created a vast network of interests all of which cumulatively eroded the subsistence system of these communities. Clearly, from their standpoint, there is a basic incompatibility between capitalist development and the realization of the oft-repeated promises of distributive justice.[37]

2. Almost every tribal community and numerous non-tribal ones have faced the violence of displacement invariably without their consent and in a manner which made them paupers within a system that made them subservient and dependent on it. Their identity, their culture and their sources of meaning were all threatened. Their collective rights have not been recognized.

3. The modern nation-state has fundamentally disrespected the distinctive cultural identities of the tribals. Both the national and internal boundaries have discriminated against them as communities belonging to the same tribe have been divided into as many as four states or between nations. This has been further compounded by a conscious programme to "integrate them into the mainstream" and that too on terms set by 'outsiders' and the state.

4. Finally, there is a shared history of state repression. Protests and movements have been crushed, activists killed, and thousands falsely implicated. Also, policies relating to forests and forest produce have reduced many a peasant or forest dweller to criminals on their own lands.

These concerns have, in different degrees and with a different sense of urgency been raised by ecologists and numerous other members of the scientific community increasingly over the past two decades. Where they differ, however, is in their prioritization of what constitutes "the environmental crisis".

There can be little doubt that this crisis, whichever way it is defined, suggests that we, as a race, have reached a turning point and that the demands that our collective economic activity are placing on the ecological resources of the planet exceed what the Earth's eco-systems can sustain. In that sense, even current strategies of achieving economic growth are going to be short-lived as they critically depend on finite natural resources while neglecting the long-term consequences of what they contribute that is in turn undermining its sustenance.

Issues Raised by People at the Grassroots

To recapitulate, the assertions of a growing number of movements and groups in the Third World are

1. Claiming greater political autonomy and redistribution of productive resources;
2. Asserting distinct cultural identities in the face of a disruption of diverse modes of social and cultural organisations and diverse systems of knowledge;
3. Asserting their separateness in ways that are having a significant political impact;
4. Challenging dominant patterns of economic development as well as the belief systems encoded in this model from the *Trumanesque conception* of development to the utilitarian justification that some must sacrifice for the benefit of others;[38]
5. Demanding greater control over productive natural resources;
6. Seeking and forming horizontal alliances within and those that transcend national boundaries; and
7. Demanding, therefore, a fundamental restructuring of the existing political and economic systems.

However, not just for these communities and their supporters, but for the planet as a whole, there are several urgent threats internal as well as external that inhibit progress towards the goal of democracy and ecological justice.

Conflicts of Interests and Conscience

Does humankind as a whole have right to the future? As so much evidence is now forcefully showing, we are losing our productive relationship with the planet itself. There are six (among numerous) important propositions and conclusions that should be considered.

1. It is imperative that we transcend economic and technocratic managerial ethic and self-consciously and politically align ourselves with those people, ecosystem and nations that have been rendered impoverished and destitute by the current patterns of economic development and ecological imperialism. Social equity is not a long-term project that can wait and social justice is not feasible without ecological sustainability. Therefore, there is an urgent need to integrate an ecological world-view into policy planning and developmental processes. In fact, unlike most Western environmentalists who neglect systemic and structural issues, there will be no Nature without social justice. The obverse is also true.[39]

2. Social justice and ecological sustainability are not feasible without centrally confronting modes of consumption. Current rates of consumption of the industrialised world and of Third World elites are incompatible with the realization of sustainability and democratic goals. Increase supply, yes; decrease waste, yes; locate and use substitutes, yes. But what of the appetite itself?

3. While in the interim, the processes of "trickle down" and of seeking accountability of state, national governments and global institutions must continue, it is imperative that greater recognition, support and legitimacy be accorded to those institutions and mechanisms that empower and facilitate relative self-

sufficiency of local communities, nations and the planet itself. One immediate and urgent task is the recovery and regeneration of degraded eco-system. Even recovering half of India's degraded lands that are capable of biotic production can support over 100 million people.

4. Multilateral institutions, like the World Bank, incapable of reform beyond a point. They are also undemocratic and unaccountable and fundamentally inconsonant with the needs of a democratic and egalitarian order. Again, while both strategies of holding them accountable as well as weakening their hold on the global and national economies must continue, the search for more accountable, more transparent and more accessible and democratic institutions must be a priority. Commitments to democracy ring shallow in the absence of this recognition. If we demand basic democratic values in our inter-personal lives we must demand it of the institutions that are defining how the planet will be organized.
5. The participation of each person in the political and economic institutions is as crucial to the future of the environment as the continuing integrity of eco-systemic processes is to the future of democratic institutions.
6. It is obvious that ecological politics emphasizes that the struggle towards an emancipatory and sustainable future is going to be a massive one, one that calls for nothing short of reversing the conquest of society by the economy; of restoring the ethic of renunciation and self-limitation; of recognising that positive values are not just material outcomes but dignity, an integration of an ecological politics and respect for the space for political struggle. Development is not feasible without listening to our conscience, without restoring ethics to economics, ethics to development, without restoring to environmentalism, an ethic and an aesthetics, without being fundamentally relevant in what we do to the last child on this planet.

Action Plan for Sustainable Development

The preparatory process for the United Nations Conference on Environmental Development (UNCED) has developed an action plan for the 1990s and beyond, commonly known as 'Agenda-21'. This Action plan forms the basis for a new global partnership for sustainable development and environmental protection everywhere in the increasingly interdependent world.

The programme of 'Agenda-21' is set out in the following facets :

1. The components broadly defined; and
2. The dominant linkages-
   (a) Cross-sectoral linkages among various components; and
   (b) The linkage between priority actions and means.

There is a clear evidence that the approach adopted by the international community to environmental rights corresponds to that applied to human rights generally. That is to be expected since the right to a healthful environment proceeds out of the international law of human rights itself. The international law system took a major turn when the human rights philosophy was made part of international jurisprudence. The sanctity attached to the doctrine of state sovereignty was diluted and human rights were accorded a super eminent status in the international system. The right of an individual to conditions conducive to his development was recognized generally, and significant areas of human rights law became accessible to intervention by foreign states.

The development of environmental law provides the second major turning point in the progress of international law. The international community has reached a stage where the individual is held entitled specifically to a natural and cultural environment ensuring the development of different dimensions of his personality. Indeed, environmental law proceeds beyond the individual to the local or national community and even beyond, to the global community, having regard to the nature and character of the environmental phenomenon. The world has entered an era in which the fundamental relationship of man to the natural order is also beginning to dawn on him.

The parallel evolution of human rights protection and environmental protection disclose some affinities that should not pass unnoticed. Both witness and precipitate the gradual erosion of the so-called domestic jurisdiction. The treatment by the state of its own nationals becomes a matter of international concern. Conservation of environment and control of pollution become likewise a matter of international concern. There occurs a process of *internationalization* of both human rights protection and environmental protection, the former as from the 1948 Universal Declaration on Human Rights, the latter some years later, as from the 1972 Stockholm Declaration on the Human Environment.

With regard to human rights protection, 18 years after the adoption of the 1948 Universal declaration, the International Bill of Human Rights was completed with the adoption of the two UN Covenants, on Civil and Political Rights (and Optional Protocol), and on Economic, Social, and Cultural Rights (1966) respectively. The normative corpus of international human rights' law is today a vast one, comprising a multiplicity of treaties and instruments, at both global and regional levels, with varying ambits of application and covering the protection of human rights of various activities.

Some international agreements like the London Ocean Dumping Convention, which controls dumping of hazardous and nuclear wastes in the marine environment; the Antarctic Treaty and the New Environmental Protocol; the World Soils Charter; the Montreal Protocol on Substance that Deplete the Ozone Layer; and the World Heritage Convention have played a significant role in shaping political dimensions of development.

The United States, which is being backed by virtually all other Western industrialized nations, lists following specific human rights to be included in the Earth Charter:

1. Right to associate with others and freely express views,
2. Right to publish and distribute information,
3. Right to participate in public debates and
4. Right to have faith and effective access to legal and administrative redress and remedy of grievances.

Therefore, political dimension of development denote a pattern which requires balanced approach in economic growth and social development because development never begins with goods, it always begins with people. Economic growth without social justice is possible but criminal, at the same time, social justice without economic growth is simply not possible, whether it is the case study of Gorakhpur or elsewhere on the planet.

NOTE

1. United Nations Conference on Environment and Development, *The Global Partnership for Environment and Development : A Guide to Agenda 21*, Geneva, UNCED, 1992.
2. For a representative critique of sustainable development from a Western Perspective, see Herman Daly and John Cobb., *For the Common Good : Redirecting the Economy Towards Community, the Environment and a Sustainable Future*, Boston, Beacon Press, 1989; Michael Redelift, *Sustainable Development: Exploring the Contradictions*, London and New York, Methuen, 1987 : and Richard Lewins and Richard Lewontin, *The Dialectical Biologist*, Cambridge, Harvard University Press, 1985.
3. For a longer treatment of the themes raised in this Article, Smitu Kothari, "Challenging Economic Hegemony: In Search of Sustainability and Justice" *Economic and Political Weekly* (forthcoming).
4. For a representative analysis of social movements in Latin America, see two books : Susan Eckstein (ed), *Power and Popular Protest : Latin American Social Movements*, Berkeley, University of California Press 1989; and Sonia Alvares and Arturo Escobar, *Latin American Social Movements*, Western Press, 1992.
5. Surajit Kumar Saha, "Historical Premises of India's Tribal Problem", *Journal of Contemporary Asia*, Vol. 16, No. 3, 1986, pp. 294-321.

6. Madhav Gadgil and Ramchandra Guha, *This Fissured Land : An Ecological History of India,* New Delhi, Oxford University Press, 1992. Also see Richard Grove, "Colonial Conservation, Ecological Hegemony and Popular Resistance: Towards a Global Synthesis" in John Mackenzie (ed) *Imperialism and the National World,* Manchester, Manchester University press 1990, pp.15-50; and Ramchandra Guha, *The Unquiet Woods: A Century of Protest in the Indian Himalaya,* New Delhi,Oxford University Press, Berkely, University of Berkely, 1990.
7. Kothari Smitu, Incompatiblity of Sustainability and Development : In Search of Social Justice, *The Indian Journal of Public Administration,* July-September 1993, Vol. XXXIX, No. 3.
8. *Ibid.*
9. The Dunkel Final Act was Distributed to GATT Delegates in Geneva in December 1991. Arthur Dunkel was till Recently the Secretary-General of GATT.
10. See in Particular, Michael Northrop, "The Uruguay Round : a GATTastrophe", *Alternatives,* Vol.18, 1993 pp.171-200. This is an excellent recent analysis of the implications of GATT on the US economy and on the range of existing standards that would inevitably have to be lowered. This may seem to be in contradiction to the intent of the US to control the GATT process. Not necessarily so since the primary actor in GATT negotiations continues to be transnational capital and more particularly the TNCs. The fact that there has been some delay in the global adoption of the GATT proposals in more a function of the differences between countries in Europe and the US on trade barriers and protectionism than the intent of transnational corporations. It is well-known that if the US and Europe were able to agree on a common set of proposals, GATT would have been almost impossible to oppose by any other part of the world. Also see Robert Weissman, "GATT's New Clothes", *Multinational Monitor,* November 1990.
11. In a way, this ideology is popularly available in Jack Hadley's novel, *Millennium* which outlines a process through which, by 21st century, the South loses control of its economy to the North.
12. Interestingly, this recognition is not new. In the 1950s, Robinson's Economics of Imperfect Competition and E. Chamberlain's Economics of Monopoly Capitalism had made the same observation though without identifying the threats to cultural and ecological pluralism.

13. Chakaravaty Raghavan, *Recolonisation GATT and the Third World*, London, Zed Books, 1991. Also see Robert Wisemen "Perlude to a New Colonisim." The Nation, March 18, 1991.

14. Malcolm Gills, "Economics, Ecology and Ethics : Mending the Broken Circle for Tropical Forests" in F. Herbert Bormann and Stephen R. Kellert (eds), *Ecology, Economics, Ethics; The Broken Circle*, New Haven, Yale University Press, 1992, p. 156. Amartya Sen has also often argued that Modern Economics has been Substantially Impoverished by the Distance that has Grown between Economics and ethics. See his *O Ethics and Economics*, New York, Basil Blackwell, 1987.

15. Joseph Petulla, *American Environmental History*, San Francisco Boyd and Fraser Publishing Co., 1977.

16. Report of the World Commission on Environment and Development, Oxford, Oxford University Press, 1987.

17. World Bank, *World Development Report*, 1991:and *World Development Report*, 1992.

18. United Nations Development Programme, *Human Development Report* 1992 and 1993, New York, Oxford University Press, 1992/1993.

19. Mahatma Gandhi as quoted in Pyarelal, *Toward New Horizons*, Ahmedabad, Navjivan Publishing House, 1959.

20. Centre for Science and Environment, *The Second Citizens Report*, 1987.

21. Michael Watts "Sustainability and Struggle Over Nature : Political Ecology or Ecological Marxism", Unpublished Manuscript, 1992.

22. Indian Space Research Centre, INSAT Data, 1991Auggest an Annual Decrease in Forest Cover of 1,3 m ha.

23. See, for example, Hewitt, *Interpretations of Calamity* or Paul Sears, *Deserts on the March*, New York Norman 1980. Extensive literature has also been generated by the United Nation Environmental Programme and by the Food and Agriculture Organisation, though the structural issues have been inadequately acknowledge.

24. Blaikie and Brookfield, *Land Degradation and Society.*

25. Bina Agarwal, *Cold Hearts and Barren slopes*, New Delhi, Allied, 1987.

26. Clarence Maloney, "Environment and Project Displacement of Population in India, Part I : Development and Deracination", *Field*

Staff Reports, Vol. 14, (1990-91).Also see Smitu Kothari, "Vikas Aur Visthapan, "*Udvasit*, Delhi. Lokayan, 1989.

27. Lokayan bulletin, Special Issue on Survival, Vol. 3, Nos. 4-5, 1984.
28. C.T. Kurien, "A Mixed Record", *India International Quarterly*, Monsoon, 1987.
29. Federation of Indias Chambers of Commerce, *Annual Report 1993*, New Delhi, FICCI.
30. Praful Bidwai, "Budget Slashes Social Sector Spending", *The Times of India*, March 5, 1992.
31. Alan. Durning, *Action at the Grassroots : Fighting Poverty and Environmental Decline* Worldwatch Paper 88, January 1989.
32. Gail Omvedt, "India's Green Movements", *Race and Class*, Vol. 28, No. 4, 1987, pp. 29-38. Also see Smitu Kothari, "Social Movements and the Redefinition of Democracy, in Phil Oldenburg (ed.) *Indian Briefing 1993*, Westview Press, 1993. For a more representative debate in India, see Jayanta Bandyopadhyaya and Vandana Shiva, "Political Economy of Ecology Movements", *Economic and Political Weekly*, Vol. 23, June 11, 1988, pp.1223-32; and Smitu Kothari and Harsh Sethi (eds) *The Non-Party Political Process: Uncertain Alternatives*, Geneva/Delhi, UNRISD/Lokayan 1985.
33. Chantal Mouffe, "Hegemony and New Political Subjects : Towards a New Concept of Democracy" in C. Nelson and L. Grossberg (eds), *Marxism and the Interpretation of Cultures*, 1988, pp. 89-104.
34. See-foot Note No.38. Also I have consciously not called these as ecological or environmental movements as this is not their self-definition of themselves. While they are attempting to integrate and address fundamental ecological issues, they are basically social mand political movements.
35. B.D.Shrama *Report of the Commissioner for Scheduled Castes and Scheduled Tribes*, 28th Report, 1986; and 29th Report, 1987-89, Faridabad, Government of India Press, 1988 and 1990.
36. Pramod Parajuli, "Power and Knowledge in Development Discourse; New Social Movements and the State in India", *International Social Science Journal*, Vol.127, pp.173-90.
37. See B.D. Sharma, "Scheduled Castes and Tribes : A Status Report", *Lokayan Bulletin*, Vol. 8, No. 6, 1990, pp. 27-38. Also see Kumar Suresh Singh, *Tribal Society in India*, New Delhi, Manohar, 1998.
38. This "trade-of" argument continues to persist. So does the regime response that these demands are not in the national interests. In

this context, significant question are being passed by movements group : Who decides what development is? At whose cost?

39. Smitu Kothari and Pramod Parajuli, "No Nature Without Social Justice", in Wolfgang Sachs (eds), *Issues in Global Justice*, London, Zed Books, 1993.

8

State Initiatives of Development

Long before India launched planned development in the traditional economic sense through a series of Five-year plans starting from 1951, Mahatma Gandhi, the Father of Nation, had perceived the concept of development in a much broader sense. Gandhiji conceived ideas for the total development of society as a whole through self-sustaining 'village republics', as he preferred to call them. With an economy that is 'biomass driven' for the vast majority of the Indian population that is rural based; sustaining this natural resource base through appropriate conservation measures has always been critical. Yet, during the last few decades, partly because of a blind faith in the Western models of development, application of inappropriate technologies, and rapidly increasing population pressure, rural development, at best can be described as only patchy.

AGRICULTURE AND RURAL ECONOMY

Agriculture is an important economic activity for a large population of the district Gorakhpur. Today, economics emphasizes agriculture and rural development as the *sine qua non* of national development. Indeed, such an reorientation, wherein rural development is well balanced with industrial growth, is now being considered essential to correct serious imbalances in the economy for equitable distribution of wealth and social justice. An estimate of population in Gorakhpur deployed in various activities can be shown with a table given as such :

Table 8.1

Year	*Percentage of workers to total population*			
	Agricultural worker	*Non-Agricultural worker*	*Total workers*	
			District	*U.P.*
1991	37.4	6.7	44.1	39.1
2001	27.4	4.9	32.3	30.9

Source : District Gazetteer.

A cursory reading of the census indicates an alarming situation in the Gorakhpur district, because of the fact that the residents here still depend mainly on agricultural occupation, living in rural areas. Though there is a decline in the percentage of population deployed in agriculture sector yet the growth of popuation in the last couple of decades has negated this declining trend. Still a large number of people are dependent on agriculture or other allied activities for their livlihood. Service sector and industurial sector are by and large absent. A large number of people involved in agricultural sector belong to scheduled castes and other backward communities (OBCs) along with their women and children as supportive hands. District administration and the State Government are therefore, required immediately to take effective and sufficient steps for rural development, poverty alleviation, SC/ST empowerment, women and child nutrition. Only then, any development can prove to be sustainable. A table of largely practised occupations in the district can be enumerated as such :

Before independence the subject of rural development and planning received little attention and whatever efforts were made in this direction were merely the outcome of political expediency and confined to such activities as sanitation and expansion of agricultural and irrigational facilities to the villages. When the first Congress government came into office in 1937, a scheme for rural development was adopted in certain villages of the district. The scope of the scheme was expanded

Table 8.2

	Number and Name of occupation	*Percentage to total workers*	*Percentage to total population*
I.	Cultivators	46.6	15.1
II.	Agricultural labourers	37.9	12.3
III.	Live-stock, forestry, fishing, hunting, plantations, orchards and allied activities	0.4	0.1
IV.	Mining and quarrying	0.0	0.0
V.	Manufacturing, processing servicing and repairs:		
	(a) Household industry	2.1	0.7
	(b) Other than household industry	2.7	0.8
VI.	Construction	0.3	0.1
VII.	Trade and commerce	2.8	0.9
VIII.	Transport, storage and communications	2.3	2.7
IX.	Other services	4.9	1.6

Source : District Gazetteer.

and a rural development association was formed at the district level, having a nominated non-official chairman and a subdivisional magistrate as secretary with functions more or less advisory. The work related largely to rural hygiene, construction of roads, establishment of libraries, construction of panchayat ghars (houses) and night schools for adults, and allied development activities. All rural development programmes were almost given up when the Congress government went out of office in 1937. In 1947, the work was again taken up when the rural development department was merged with the co-operative department and the rural development association replaced by the district development association with a non-official as chairman and the district co-operative officer as its secretary.

The development activities received proper attention only after the setting up of the Planning Commission of India in 1950. In the First Five-Year Plan (1951-56), agriculture, including irrigation and power, were given top priority, funds having been allocated accordingly. In 1952, the district development association was replaced by the district planning committee, having the district magistrate as its chairman and the district planning officer as its secretary. Its role was, however, advisory. A number of sub-committees were also framed for the preparation and execution of development programmes.

In the district the first community development block was opened on October 2, 1953, at Nichlaul followed by another at Chargawan on January 26, 1955.

The scope of the Second Five-Year Plan was enlarged to include industrialization and it was decided that the whole district would be divided into development blocks for the implementation of the various Plan schemes. In the second half of the Second Five-Year Plan the Antarim Zila Parishad (Now the Zila Panchayat) was formed in 1958 by amalgamating the District Planning Committee and the District Board. A three-tier structure of rural self-governing bodies was adopted from December 1973, for the successful implementation of the planning and development programmes–the village panchayat at the village level, the Kshetra Samiti at block level and the Zila Parishad at the district level. For the co-ordinated execution of the different Plan schemes the resources of the agriculture, co-operative, animal husbandry, Panchayati Raj and some other departments and organizations were pooled and put under the control of the district planning officer (now designated as district development officer). The district was divided into 31 development blocks.

The Kshetra Samiti is responsible for all the development activities within a block. The block development officer is the executive officer of the Kshetra Samiti. He is assisted by a number of assistant development officers for agriculture, animal husbandry, co-operatives, panchayats, etc. At the village level, there is a multipurpose worker designated as Gram Sewak (village level worker, now designated as Village Development Officer) to work for all the development departments.

During the First Five-Year Plan period the stress was on people's participation in different activities. Improvement of village roads, drainage and pavements were done by Shramdan (voluntary labour). Improved methods of agriculture and use of compost manure were also introduced. Tube-wells and other facilities for irrigation were augmented.

The aim of Second Plan was to increase national income by 25 percent and to reduce unemployment. During this period branches of different co-operative societies were established to enable the farmers to get proper value of their produce and a land development bank also started functioning, to make available to farmers loans on easy terms. In the field of agriculture, schemes relating to soil conservation, Japanese method of paddy cultivation, expansion of and training in, the use of improved agricultural implements, and use of chemical and green manures were taken up.

The Third Five-Year Plan was conceived as 'the first stage of a decade or more of intensive development leading to a self-reliant and self-generating economy.' It sought to ensure a minimum level of living to every family while narrowing economic and social disparities. Some special programmes, such as intensive methods of wheat and paddy cultivation, crop protection measures, availability of improved varieties of seeds, increase in double-cropped areas, were taken in hands. This, all resulted in substantial increase in production. Schemes for the development of livestock, poultry and fisheries were also undertaken.

A period of three annual plans intervened between the Third Plan and the Fourth Plan, which commenced from April 1, 1966, the broad objectives of these Plans being the following:

(i) A growth rate of 5 percent in the agricultural sector and 8 to 10 percent in industry,

(ii) 6.9 per cent annual rate of growth in food-grains production to achieve self-sufficiency,

(iii) To maximise employment opportunities with the proposed growth rate,

(iv) To redress imbalances arising from a high rate of population growth and inadequate expansion in agricultural production by bringing the fertility rate to 25 per thousand in shortest possible time.

The implementation of these Five-year Plan schemes has helped in raising the standard of living, providing better wages and living conditions and helping the general economic growth of the district. The growing industrialization side by side with increased agricultural production holds a definite promise of ushering in an era of increased prosperity and better economic condition for the people of the district, including SCs/STs. This will not only improve their financial conditions but will also make them politically aware and stronger. The road to success in society and polity always go along with financial success.

Marginalisation of Farming Communities

A very large section of the farming community in Gorakhpur has remained unaffected by modern, high-energy input, agriculture technology, partly because of limited availability of energy subsidies and partly because of economic reasons. Indeed, increasing proportions of our rural communities have got marginalized. The worst affected are the entire SC/ST population residing in pockets in different parts, though more often these pockets are richer in natural resources. Other traditional societies, such as, backward classes in the entire Gorakhpur district have also not had any access to innovations in agricultural technology. Not, only these communities have been adversely affected by the activities of sugar cane industries in the entire eastern Uttar Pradesh, having over-exploited the natural resources in their territories in the name of regional development, their problems have been accentuated due to increasing population pressure and the consequent fragmentation of the land holdings, now impoverished by environmental degradation. Wherever alternate technologies have been identified for them, they invariably have been totally divorced from their traditional value system, and the identified technologies are often economically non-viable and ecologically not sound. It is in this contest that we looked at the whole question of sustainable development of the scheduled castes and other backword communities of the eastern Uttar Pradesh in general and that of Gorakhpur district in particular, by building upon traditional technology as the starting point.

Equity should be as much a corporate goal as is efficiency. The environmental community, too, must face up to the reality of environmental inequality based on caste and income. If environmental justice is to be achieved, it can only be on the basis of being indivisible as to caste and income.

Numerous studies dating back to the early 1970s document that lower-income persons, working-class individuals, and people of SC/ST category are exposed to elevated health risks in their homes, in their communities, and on their jobs. People of SC/ST communities have borne greater health and environmental risk burdens than the society at large. This is the case even when income is held constant. For example, community has been found to be independent of class in the distribution of air pollution, contaminated fish consumption, location of municipal landfills and incinerators, abandoned toxic waste dumps and fluoride poisoning in children specially, in the rural areas of Gorakhpur.

Government have been slow to address environmental justice and equity questions of who gets what, why, and how much, who can afford help and who cannot, why some communities are poisoned and others escape, and why some populations are studied while others get left off the research agenda. Today, disenfranchised communities are calling for greater input into shaping environmental priorities that were in the past, left to a small homogeneous group of scientists, health professionals, government bureaucrats, and environmentalists.

The impetus for getting environmental justice concerns on the national agenda has not come from regulatory agencies or the national environmental groups. The impetus has come from the environmental justice movement. This movement is comprised of a loose alliance of grassroots and national environmental activists, civil rights and social justice leaders, and academicians who question the foundation of the current environmental protection paradigm.

The current environmental protection paradigm manages, regulates, and distributes risk. In combination with systemic casteism, and a general class bias, the current system:

1. Institutionalises unequal enforcement;
2. Trades human health for profit;
3. Places the burden of proof on the "victims" and not the polluting industries;
4. Legitimises human exposure to harmful chemicals, pesticides, and hazardous substances;
5. Promotes "risky" technologies;
6. Exploits the vulnerability of economically and politically disenfranchised communities;
7. Subsidizes ecological destruction;
8. Creates an industry around risk assessment;
9. Delays cleanup actions; and
10. Fails to develop pollution prevention as the dominant strategy.

Sustainable Agriculture and Rural Development

Hunger and malnutrition are endemic among developing countries. By the year 2020 about 84 percent of the expected global population of 8.5 billion will be living in developing countries. The fundamental challenge facing agriculture in developing countries today is to increase food production in a sustainable way and feed expanding populations. Such an increase has to come primarily through intensification of current agricultural production, as the potential for bringing new land under cultivation in many countries is very limited. If hunger is to be eventually eradicated in these countries, this intensification must be both ecologically and socio-economically sustainable.

Agriculture is an important sector, if not the backbone, of the national economies of many developing countries like India, sometimes representing the major share of export earnings. Over the next 10 years this sector will bear most of the responsibility for providing rural economies with sufficient growth to offer employment and other remunerative activities to the bulk of their population, even in the most backward areas of the country, such as eastern Uttar Pradesh. Without this growth, the present rural-urban exodus is certain to accelerate, leading to unmanageable urban squalor and, in all likelihood, major socio-political upheavals.

Major adjustments are needed in national and international policies to create conditions for sustainable agriculture and rural development. A major objective is to increase food production in a sustainable way and enhance household food security. This involves educational initiatives, utilization of economic incentives and the development of appropriate and new technologies, thus ensuring stable supplies of adequate food by vulnerable groups and production for markets, employment and rural income generation as well as efficient natural resource management and environmental protection. Priority must be given to maintaining and improving the capacity of higher potential agricultural lands to support an expanding population. At the same time conserving and rehabilitating lower potential land is essential. Emphasis should also be given to reforming agricultural policies relating to price, subsidies, trade, land tenure and appropriate farm practices and technologies.

In promoting sustainable agricultural and rural development, the commitment of national governments and the support of the international community in implementing the programme is crucial. Efficient agriculture needs to be job-creating, both on and off farm and thus help reduce poverty levels. Emphasis should be given to reform agricultural policies relating to price, subsidies, trade, land tenure, and appropriate farm practices and technologies. Major investments in rural infrastructure will be required in promoting sustainable agricultural and rural development; the support of national governments and the international community in implementing the following action programme is crucial.

An action programme on *land resources planning information and education for agriculture* should be developed to systematically identify sustainable land uses and production systems for each land and climate zone, to control inappropriate land use and take into account the actual potential, carrying capacities, and limitation of land resources. In addition, a set of actions on *sustainable plant nutrition to increase food production* will formulate strategies to maintain soil productivity. In many developing countries, where population growth exceeds agricultural production, the goal is to increase agricultural production substantially, particularly in high potential areas, without destroying soil fertility. This will require determining and optimizing the use of organic fertilizer and other sources to enrich soil and thus increase farming efficiency and production. This requires empowerment of SC/ST communities on which depends the agricultural sector of the district. The state must transfer adequate financial and administrative powers to the local bodies and Panchayati Raj Institutions, in order to make policy and programme implementation effective.

Another key programme is *improving farm production and farming system through diversification of farms and non-farm employment and infrastructure development.* Intensification should emphasise farm management technologies such as crop rotation, organic manuring, efficient and economically viable use of chemical inputs, agricultural waste recycling, and the prevention of pre-and post-harvest losses. Where intensification is not possible, other employment opportunities, such as private small-scale agro-processing units, cottage industries, wildlife utilization, fisheries, conservation and reclaimation activities, and rural infrastructure development should be created.

District administration of Gorakhpur should also implement policies pertaining to land tenure, demographic trends, appropriate farm technologies and a more open trading system that would enhance rural households' access to food.

Regional and sub regional cooperation is essential for community mobilization. The regional commissions, regional development banks and regional economic and technical cooperation organizations, within their respective agreed mandates, can contribute to this process by:

1. Promoting regional and sub regional capacity-building;
2. Promoting the integration of environmental concerns in regional and sub regional development policies;
3. Promoting regional and sub regional cooperation, where appropriated, regarding trans-boundary issues relate to sustainable development.

There must be active cooperation and collaboration among the regional commissions and other relevant organizations, regional development banks, non-governmental organizations and other institutions at the regional level. UNEP and UNDP, together with the regional commission, would have a crucial role to play, especially in providing the necessary assistance, with particular emphasis on building and strengthening the national capacity of Member States.

REORGANISING GRASSROOTS INSTITUTIONS FOR SUSTAINABLE DEVELOPMENT

During the last five decades of planning, the country has witnessed spectacular success in various sectors of economy. This became possible mainly due to a number of programmes introduced from time to time embracing different sectors of our economy. The increased use of improved crop cultivation and fertilizer along with huge investments in inflation considerably helped in improving agricultural scenario of the country. In addition, improvement in infrastructure and institutional support also contributed to this development. The number of irrigation pump sets has increased from 200 to about 7,000 in Gorakhpur within five decades of planned development, and power generation from a little over 1,000 MW in 1947 to 50,000MW in 2001 and fertilizer production has increased from a mere 18,000 tonnes to 2,20,000 tonnes. As a result of the efforts made by the government, the district moved ahead and the people strove to achieve, in a rather short span, what had not been achieved all along. These efforts laid the foundations of rapid agriculture and industrial growth, and improvement in social services such as education and health to keep pace with the ever-increasing population.

Planning and development have demonstrated a strong correlation between agricultural growth and economic prosperity. Although, agricultural development in the whole country, not in Gorakhpur alone, is characterized by success in self-sufficiency, yet it gives a picture of a mix of great achievement and opportunities. India has about 50 per cent land suitable for cultivation compared to global rate of 10 percent only. Despite substantial investment in irrigation, its potential is not being fully utilized and yields are extremely low in comparison to the best in the world.

A large segment of rural population in Gorakhpur is poor. According to various estimates, the incidence of poverty in rural areas varied between 38 and 63 percent in 1967-68, depending on the various definitions of poverty line adopted in the respective studies. For instance, according to the census of 2001, out of the total district population of 37,84,720, about 40 percent is living below the poverty line. A majority of rural people belong to agriculture labour and household, both with and without land as well as:

1. Other landless rural household; and
2. Marginal and small farmers, operating less than five acres of land.

The weakest sections of population are mostly the agricultural labourers, generally comprising of Scheduled Castes (SCs) and Scheduled Tribes (STs). Both these groups together constitute about 79 per cent of the total rural population.

In the past, a number of steps have been taken to alleviate poverty and improve the standard of living of especially the weaker sections of society. In addition to several region-specific area development programmes such as SFDA, MFAL, ICDS,

DPAP, DWCRA, IRDP, TRYSEM, JRY, etc, for different target groups were also initiated with twin objectives of increasing income level of rural people and minimizing inequality in income distribution.

The experience indicates that despite initiatives taken by the government to help rural poor and reduce incidence of poverty, many of the programmes could not yield desired results mainly due to lack of people's participation and well organized institutional support, particularly at the grassroot level. Therefore, a number of measures were taken to ensure that the benefit of these special schemes reach the real poor. For instance, people belonging to weaker sections were given priority in the development efforts so that the benefits of planned investment can go to the relatively backward sections of the community.

People and resources are the two key elements of development process and unless there is a balance between these two, any development will not be sustainable. The district population has been growing at the rate of 23.44 percent per decade, of which 64.78 percent comes mainly from rural areas. Since, agriculture not only supplies raw material for some of the major industries, such as Cotton, Jute, Textile and Sugar but also contribute substantially to the state's earnings. Increasing population might adversely affect the production of these crops because of ever-increasing demand for food. If the population growth is not contained, it might put pressure on natural resources and hamper the process of economic growth. Excessive use of these resources, without giving sufficient time for regeneration of their productive capacity, will lead to environmental degradation, making overall economic development unsustainable.

Poverty is often indicated as a reason for environmental degradation. The poor people are both agents and victims of environmental degradation. However, the main cause of rural poverty leading to environmental degradation is abundant use of traditional practices, collapse of local institutions, deforestation, and soil erosion in small farmers' field. Therefore, the need for mutually reinforcing packages of technology, services and government policies for maintaining the momentum of progress has been rightly advocated. It has also been stressed that without land reforms, a long-term interest in strengthening the production and infrastructure of the farm will be lacking.[2] It will require a sustainable and strong network of local institutions to effectively help in implementing these development schemes.

Agriculture growth must not only contribute to the overall growth of the economy but it must be sustainable as well. The following shortcomings in agricultural development have generally been indicated in the district Gorakhpur as such:

1. Poor resource-base in agriculture,
2. High rate of growth of rural population,
3. Low productivity in agriculture, and
4. Poor infrastructure.

Obviously, the starting point has to be full employment or near-full employment. Our experience of nearly four decades of planning clearly shows that employment cannot be just a by-product but will have to be planned on a massive scale, starting with the unskilled labour force in rural areas and those who can be employed in agriculture and allied activities, simultaneously planning to impart training for a variety of skills that would be needed in a developing economy. This approach will require an in-built mechanism to sustainable development. It will however, also require appropriate linkage with local-level institutions so that a wholesome mobilization process could be initiated.

The mid-term appraisal of Seventh Five- year Plan also emphasized the involvement of Panchayati Raj Institutions in development process as follows:

"However it is noticed that wherever the Panchayati Raj Institutions have been actively involved, the implementations of rural development programmes have been better and the selection of beneficiaries and designing of schemes have been more satisfactory. The Planning Commission has been impressing upon the states that various rural development programmes will be realistic and meaningful only if people's representatives are actively involved and associated in local level planning, designing, formulation and implementation of those programmes, and selection of beneficiaries in the anti-poverty and employment programmes such as IRDP, NREP, RLEP, etc., and there is no better instrument to meet this need other than the Panchayati Raj Institutions.[3]

The Renewal

However, despite the emphasis on active role to be played by Panchayati Raj Institutions, they have not been that much effective.[4] Hence, a number of efforts have been undertaken to bring forward a Constitutional Amendment Bill 1992 to provide constitutional status to the Panchayati Raj bodies and to strengthen them by making suitable provisions for durability; continuity; adequate representation to SCs, STs and women; financial allocations; and regular elections.

The 73rd Constitutional Amendment came into being keeping in view the following objectives:

1. To place more and more powers in the hands of the rural people to determine their own destiny and to build a society;
2. To enhance the capabilities of the rural people to frame planning from below;
3. To decentralize execution of all kinds of developmental activities with effective participation of people; and
4. To alter the administrative methods at rural lands through Panchayati Raj.

The Amendment provides the Legislature of a state to authorize the panchayats to utilize powers, authority and responsibilities with respect to:

1. Preparation of plans for economic development and social justice, and
2. Implementation of schemes for economic development and social justice.

Though a set of 29 items specified in the Eleventh Schedule have been entrusted to the panchayats, but in addition any other schemes for economic development and social justice may also be entrusted to them.

The Constitution Amendment on Panchayati Raj thus attempts not only political decentralization but also representation of poor and underemployed sections of the society in local institutions. The SC and ST communities account for a substantial part of the landless labourers and rural artisans. Democratic decentralization of administration and development, as a result, will reach down to the *gram sabha* and the panchayats, at the village, at the intermediate, and at the district levels. Additional powers are vested in elected representatives of the rural people to determine their priorities and to build a society at the village level, where the process of sustainable development shall be started by them for themselves through the Panchayati Raj Institutions. The rural people will get an opportunity to enhance their capabilities to frame plans form below, to execute all kinds of developmental activities with effective participation of people and to alter the administrative methods at local levels through the grassroot institutions. The development programmes, in general, and the Jawahar Rozgar Yojana, in particular, have got a boost as a result of this Amendment. The basic principle of the Jawahar Rozgar Yojana is that the Gram Panchayats, which get an assured allotment under the programme, will decide on the schemes to be taken up in their villages. With an elected body at the panchayat level responsible to the electorate and the will of the people, will find greater reflection in the implementation of the Jawahar Rozgar Yojana and other rural development programmes.

Keeping in view the fact that agricultural development in the country can no longer be considered independent of increasing population pressure, on the one hand, and conservation of natural resources, on the other, strategies for institution building process at the grassroot level will have to be given prime place. In a vast developing country like India, where more than one-third of the people are still below the poverty line, the benefits of developments do not automatically trickle down to the poor but have to be taken to them in the form of education, health, social services, employment and skill. This is where the institutional apparatus and the administrative system will have to intervene and interact with the poorest of the poor, to enable them to develop faster and build decentralized institutions, which will contribute to sustainable development.

NOTE

1. P.S. Ramakrishnan, *Shifting Agriculture and Sustainable Development of North-Eastern India*. UNESCO-MAB Series, Paris, Parthenon Publishing, Carnforth, Lanes, U,K., p. 424 (republished by Oxford University Press,New Delhi, 1993).
2. M.Y. Ghorpade, Development Ethos and Experience, Bangalore, *Southern Economist*, 1991, p. 12.
3. Government of India, Planning Commission, "Involvement of People in the Planning Process", Paper Placed before the Meeting of the Consultative Committee of Parliament, June 30, 1987, Delhi, Manager Publication, 1987, p.16.
4. S.P. Jain, "Panchayati Raj and Agriculture Development : Some Issues", *Indian Journal of Public Administration*, Vol. XXIII, No. 3, July-September, 1977, pp. 721.

9

People's Participation in Sustainable Development : A Critical Analysis

Government agencies, over the past 50 years or more, have tried in vain to change methodology of agriculture, which demands high energy inputs in the form of fertilizers, weedicides and pesticides. As the soil cover is thin and infertile and the nutrient losses from the system are very heavy, more and more fertilizer is required to sustain the system over a time period, but with very low efficiencies. As weed problem gets exaggerated under sedentary farming, weed control assumes alarming proportion. For these ecological reasons and for a variety of social and cultural reasons related to land tenure and cultural and religious practices centered around traditional agriculture, the farmer rejected the alternatives to shifting agriculture as a permanent solution to this problem.

It is in this context that we stepped into the region to look at the whole issue of sustainable development, with people's participation. A holistic approach for sustainable development that would link up agriculture, animal husbandry, and domestic sub-system of the village eco-system in the overall context of forest eco-system function and management was identified.

The whole approach was to build upon traditional technology and knowledge through modern scientific inputs, based on a value system with which the people can identify themselves and therefore participate effectively in the developmental process.

Specially designed packages have to be developed for each cluster of villages, taking account of micro-climatic conditions, socio-economic levels, and socio-cultural background of the people. Scientists, planners and administrators have often tried to impose development plans that they consider are good for people in the region, without trying to understand the social processes that operate in traditional societies and the linkages with eco-system level processes. Planning should try to link up ecological processes within a landscape (a set of interacting eco-system types), with social processes operating within the system. Such a planning philosophy would take care of the traditional value systems, and therefore, would not only find ready acceptance by the traditional societies but would also ensure their participation in the developmental process. For instance, the village headman or Gram Pradhan, as popularly called in the Gorakhpur district, is the key person in each village panchayat. Village level institutions linked with the District Council infrastructure could provide the necessary institutional framework for development planning. The planning approach should incorporate a review period so that appropriate adjustments may be made based on experience.

Institutional Framework

Building upon traditional knowledge and technology for land-use in the eastern Uttar Pradesh implied that by basing development strategies on a value system which people understand and appreciate and therefore they are able to identify themselves with it, peoples' participation, would have to be ensured right form the formulation stage. Building village level organizations are critical in such an effort for sustainable development.

Indeed, an important indicator of sustainable development is to look at the local level institutional framework considering the following aspects:

1. Identification and strengthening of local level institutions that are already available such as those existing in the northeastern region;
2. The representative nature of these bodies and the extent to which individual's or family's interests are taken care of;
3. Their role in decision-making right from the project formulation stage through different levels of implementation;
4. Flexibility in function so as to take care of the interests of all sections of the society;
5. Education and human resource development that these institutions have been able to trigger, particularly for weaker and vulnerable sections of the society; and
6. Ability of these institutions to stand on their own through empowerment in terms of capability building.

WHERE DO WE GO?

There are a number of good examples of efforts made to implement sustainable development at local level with people's participation. The Badhvi Tal Development Programme near village Dhani (now in district Maharajganj) and the Forest Development Committees formed between the forest dwellers and State Forest Department of the Gorakhpur district are two of the examples where village level institutions have worked well, with some efforts already put in for replications of these experiments. However, the major deficiency in these and many other such voluntary group efforts is that rigorous eco-social analysis is wanting, and where available they are of all limited nature. In the development block Sahjanwa of the district Gorakhpur, the experiment with a nitrogen fixing tree, *Nepalese alder*, as an attempt to redevelop shifting agriculture *(Jhum)* affected land-use, based on empirical knowledge already available with local scheduled caste societies in the region, and further refined by rigorous agro-forestry inputs, is one example where practical field based research analysis and development have been effectively blended.[1]

COMMUNITY PARTICIPATION FOR SUSTAINABLE DEVELOPMENT

A large number of projects are being implemented both by the government and non-governmental organizations in various sectors, *viz*, rural development, health, family welfare, education, urban development, etc., which are either funded by the local resources (Central and State Governments) or by the grants from foreign donor agencies. The future of these projects becomes uncertain when the funds are withdrawn after the prescribed period. In such cases, there is no continuous flow of benefits to the beneficiaries. Most of the projects vanish when the funds and personnel are withdrawn. In short, it may be said, that the projects spring up as soon as the funds and personnel are available by the donor agencies and the projects cease to exist once the agencies withdraw their support. If development is one-time procedure, then the question of continuity of efforts or sustainability would not arise. But it is not so. Hence, there is a need to think of means of generating self-sustaining efforts so that the projects once initiated would go along and produce continuous benefits. Therefore, effective and sufficient participation of people is essentially required.

The consensus is that if development is to be sustainable, it must be participatory and community based. The development initiatives should be based on the needs identified by local people, involve them in the designing and implementation of the projects, using principles and techniques suited to local conditions. This process would include all sections of the community, especially women.

Development is for people, and as such, an essential element in the development process is people's participation. Development cannot be imposed from outside or above, the desire for development must come from within or from those who need to be raised from the lower rungs of economic/social ladder to higher rungs. It may be argued that in some instances, people would preserve the *status quo* instead of change in the situation to which they have been accustomed. People can be made aware of their needs to realize their resources and potentialities and to participate in development process.[2] Rembember, development never begins with goods, it always begins with people.

Thus, development may be conceived as improving living standards of the masses of the low-income population residing in rural areas and making the process of their development self-sustaining. One of the important elements is community participation. The Fifth International Action for Development Conference stressed that development can be a reality with the involvement of people only.[3] People should be regarded as subjects and not just as objects of development. The emphasis should be on the premise that the projects are for people and not people for projects. To involve people in development is a slow process, but once set in motion, it can produce amazing results.

The concept of participation is not something new. The ancient Greek Scholar, Aristotle found a clear relationship between the extent of participation and the creation of a good life. According to him, the best state was one where there was broad participation with no class dominating the others. His analysis showed some relationship between the participation and development.

One school of thought equals participation with democratic values and believes that participation will produce many additional positive results.[4] If people take active role in solving their own problems and meeting their own needs, they will acquire the power that was previously retained by government by default.

Development professionals have advocated a school of thought, which urged a humanistic-democratic participative management philosophy as the only practical way to get the beneficiaries committed to the project and to build local capacity.[5]

A group of development administrators has advocated participation essential to the sustainability of projects.[6] Oakely is of the view that till now development planners had overlooked the contributions that people could make and the skill that they could bring to development project.[7] If the project planners and administrators could incorporate the human element in such project and persuade the people to participate in them, then there would be a stronger chance that these projects would be successful and self-sustaining.

PARTICIPATION : IT'S MEANING

The term participation has its origin from Latin word 'participare', which means 'taking part'. Encyclopedia of Psychology describes participation as;

1. "Taking part or involvement in an activity', and
2. "Greater involvement of person in policy decision which affect them directly".[8] In its simplest term it connotes self-activity and is a product of collective action in relation to social and economic situation.

Cohen and Uphoff regarded participation as "generally denoting the involvement of a significant number of persons in situations or actions which enhance their well-being, *e.g.*, their income, security or self-esteem".[9] Paul defined participation as, "in the context of development, community participation refers to an active process whereby beneficiaries influence the direction and execution of development projects rather than merely receive a share of project benefits".[10]

This implies that for effective participation, leading to sustainability of projects, people of all class and caste should be involved in all stages of plan process. The dream of participation will become a reality only when the 'top-down' approach, which is prevalent now, is replaced by 'bottom-up' approach.

Lele, after examining the nexus between the large number of factors and the capacity of the project to carry out the rural development tasks, concluded that local participation may mean involvement of people in planning, including assessment of local needs. Even if the people do not participate in planning, at very minimum, they should be informed of the plans designed for their areas if they are expected to give consent and to cooperate with the programme implementation. Participation in planning and implementation of programmes can develop self-reliance, which is necessary among rural people for accelerated development. The situation, as it obtains now, is that people in poor communities are the torpid recipients of 'improvements', which had little connection with their problems as they saw them.

The ILO points out that "participation involved active, collectively organized and continued efforts by the people themselves in setting goals, pooling resources together and taking actions which aim at improving their living conditions.[12]

ACC Task Force on Rural Development views participation in terms of 'active' and 'passive' process. It points out that "what gives real meaning to popular participation is the collective efforts by the people concerned to pool their efforts and whatever other resources they decide to pool together, to attain the objective they set for themselves".[13] In this regard, participation is viewed as an active process. The idea of 'passive' participation involves the people in action that have been thought out, designed and controlled by others.

An analysis of all the available definitions or descriptions of participations indicates that experts/researchers defined the term to suit their studies. This shows that because of its complexity in nature with varied dimensions, it will be difficult to contain participation in any formal definition. However, based on these diversities, the researcher has identified some commonalities among these definitions to arrive at an understanding of the concept. To effectively participate, persons/group has to identify themselves with the movement. This would be possible only when people are involved in need identification, fixing priorities among the needs, and associated in the planning, decision-making, implementing and evaluation process and given an opportunity to create/maintain/manage local organizations. In general, it can be said that community participation implies sustainable development with and by the people and *not just for* the people.

FACTORS INFLUENCING PARTICIPATION

Majority of the respondents expressed that possible ways of involving people in development activities are:

1. Convincing the people about the importance of the programme;
2. Need for better planning with people;
3. Need to consult the local organizations before launching the programme;
4. Inviting the villagers for discussion before introducing the programme; and
5. Taking into consideration the needs of the people.

Another important factor is awareness. The extent of participation depends on the degree and level of knowledge the people have regarding the programme. In the absence of clear, correct and complete information, the beneficiaries are not in a position to take any decision to participate or otherwise. The doubt creates fear and hence they prefer to remain passive spectators of various schemes.

The development planners and administrators need to understand the cognitive map of rural people, the way they perceive the programme, its benefits, costs and risks. The people would decide to participate depending on how much it would contribute to their goals and then the larger interest of the community. They would compare the costs that have to be incurred and the relative gain, which would accrue to them and to community before they decide to participate. Based on this assumption, Bryant and White[14] proposed the following formula:

$$P = (B \times Pr) - C$$

Where	P=	Participation which depends on
	B=	Benefits which one hopes to get,
	Pr=	Probability that they will actually achieve the benefits, and
	C=	Costs involved in work for the projects.

This formula predicts that people will participate in an activity when they sense that their benefits outweigh the time and effort they are expected to spend. It should not be constructed that this formula would make the trick of making people to participate. However, this provides insight into formulating certain basic principles of participation (modified from Gow and VanSant), which are:[15]

1. People organize best around the problems they consider most important in assessing needs and in planning initiatives to meet those needs.
2. Local people make rational economic decisions in the context of their own environment and circumstance.
3. Voluntary commitment of labour, time, material and money to a project is both an evidence of participation and a necessary condition for breaking patterns of development paternalism.
4. Local control over the amount, quality and distribution of benefits increases the chances of the project becoming self-sustainable.

The above-mentioned premises give certain clues for identifying the following approaches which one might adopt possibly to enlist community participation for sustainability:

1. People should be made aware of the details of all the schemes.
2. Identification of felt needs of the people/community.
3. The planning process should be of "bottom-up" approach. The user of the programme should also be the decision maker in the programme. The present "top-down" approach should be changed to another approach called "bottom-up with support-down", where the rural people plan their programme with the official support.
4. Developing a sense of ownership and responsibility for a programme among the community members to create a feeling of possessiveness. To develop this, the outside

agency should be prepared to gradually withdraw its role, delegating the responsibility and functions to the community.

5. Subsidy should be given only to serve as supplement to the existing resources of the individual/community. A subsidy is most effective, if the individual/community takes it as a form of recognition of the effort and that it is aimed at stimulating more effective and sustaining development.

SALIENT RECOMMENDATION OF YASHADA WORKSHOP

During February 1991, Yashwantrao Chavan Academy of Development Administration (YASHADA), Pune, organized a two-day workshop on sustainable development. Senior administrators, programmers and academicians after deliberations came out with certain recommendations.[16] Some of them which have direct relevance to the theme of the research are given below:

1. Government initiatives at village levels should involve more policy participation and decision-making from villagers instead of restricting it to institutional mechanism to the panchayats.

2. As recommended by Shri P.B. Patil Committee report, local administrative bodies, like Zilla Parishads, Village Panchayats and Panchayat Samities, need to be immediately strengthened to ensure effective decentralization and optimal resource management.

3. It was recommended that less of subsidized schemes should be introduced and that subsidies to be withdrawn in phased manner.

4. The government may be persuaded to set up an institutional mechanism to review and approve innovative ideas that promote sustainable development, give access to policy makers and ensure reduced reaction time, to immediately implement these ideas.

5. Norms of sustainability should be introduced while formulation and designing projects.
6. Minor irrigation projects and participatory watershed development projects should be encouraged. Canal committees of people may be set up for effective water management and for minimizing water losses.

Development, being a multi-dimensional phenomenon, requires effective involvement of the community in achieving its goals. In a developing country like ours, the resources for development at the command of the government are very little and scarce. As such, in order to carry out local sustainable development measures effectively, community needs to be mobilized-physically, psychologically, financially and materially—to supplement the resources provided by the government and other extra-community sources. In overall strategy of sustainable development policies and implementation work plans, community inputs in the form of systematic arrangement for participation of the people of the community in policy making and for a regular feedback in regard to sustainable development programmes may help in decision-making process at different levels. The community could raise substantial financial and human resources from the community for sustainable development which otherwise might remain unused such as the enthusiasm and energy of the youth of the local community and women of the area. Instead of relying upon voluntary and individual contributions, the community with institutional structures- such as local self-government, cooperative societies, etc., having a control over the local community could mobilize resources for community's social welfare purposes more easily. Therefore, the community participation is most widely visualized as a means to mobilize resources for sustainable development which otherwise might not be available. Inspite of its sincere and best efforts, the government cannot promote sustainable development entirely from its own resources. It is, therefore, desirable and imperative that the intended beneficiaries themselves contribute to the extension of sustainable development, just to minimize the cost of the development projects.

Purpose of Participation

Based on our above statement, it becomes imperative to examine the purpose which participation could achieve. They are as follows:

1. The most important purpose of participation is to create will and determination among the members of the community to improve upon their present and future life.
2. To identify and develop the local resources to generate self-reliance among the members of the community.
3. To achieve integrated area development by coordinating among various agencies interested in sustainable development.
4. To mobilize available human resources for productive and meaningful activities.
5. Being in constant touch to make the members aware of the community about the development in the area.
6. To help the community to understand the new technology by way of arranging functional literacy programmes.
7. To create centres for discussion and development by organizing various clubs of youths and women.
8. To provide an open forum to the members of the community where they could discuss their problems and find indigenous solutions, which may prove, efficient and economical.
9. For continuous education and mobilization of the people in the area to develop local leadership.
10. For accelerating socio-economic development of the community to encourage people to adopt modern changes.
11. To arrange extra-curricular activities for generating social-awareness by the way of well-designed publicity.

12. To encourage people to develop their own confidence instead of depending upon the government, and become self-reliant, which is key to development.

To improve the efficiency and efficacy of sustainable development programmes, the hand in glove relationships between the people and development organizations may be beneficial for both.

The Fifth Five-Year Plan recognized it well when it stated, "the involvement of the people and their elected representatives is a pre-requisite for effective planning. A plan which does not take into account their aspirations and preferences can have no operational validity, especially since its successful implementation can be ensured only if the majority of citizens functioning as entrepreneurs and decision-makers in relation to consumption, savings, investment, etc., endorse the envisaged policies and programmes, by that wholehearted participation is ensured". More emphatically this philosophy was retained in the Seventh Plan document.

Process of Participation

Since the community consists of many heterogeneous groups, the understanding about the community itself becomes the first pre-requisite of participative management. It helps in understanding the positive as well as negative attitude of the community towards a particular set of activities and ultimately helps in accelerating the process of participation. However, this exercise must be carried on scientific lines and attempt should be made to locate informal leaders within the community through the process of sociometry. If such informal leaders are located, they may help in changing the heterogeneous character of the community in homogeneous one and may help in avoiding the wastage that results from the friction among different groups.

The identification of the priority needs of the community may be treated as second pre-requisite and the development workers in the area must design the development programme to meet the identified and realistic needs of the community. At the stage of identification of community needs, it is desirable that the opinion leaders- both formal and informal- are involved in such exercise. It may help in obtaining such information, which might not have been collected during the formal survey. It may also help in minimizing the clash of interests.

After identification of needs, when the programmes of sustainable development are designed and got approved by the headquarters, the development workers should educate the people both through formal and informal channels about the usefulness and potentialities of the programme. At this stage, development workers are also required to assess the resources, which could be tapped locally, involving no cost or very little cost, so that people could manage the development programme on their own at some later stage.

The leadership role of the development workers is one of the most important steps in the process of participative management. However, the development workers can play their crucial role in participative management only when they are adequately and properly trained in science and arts of management. Thus, the success or failure of participative management would largely depend on administrative capability, and leadership and motivation of development workers.

When the sustainable development programmes take-off with aided guidance of bureaucracy, it becomes important for development workers to encourage the people to run the programmes independently without their help. This may repose confidence among the members of the community and bring a sense of confidence and responsibility to take up the development measures independently, which is key to the participative management. However, it is expected that members of the community may face some problems in initial stages and as such the development workers are required to occasionally monitor the running of the programme and if needed, solve the problems faced by the community. Such types of problems may be varied and many, *i.e.*, political, religious, social, economic, psychological, administrative and so on. The

Chart : 9.1[17]

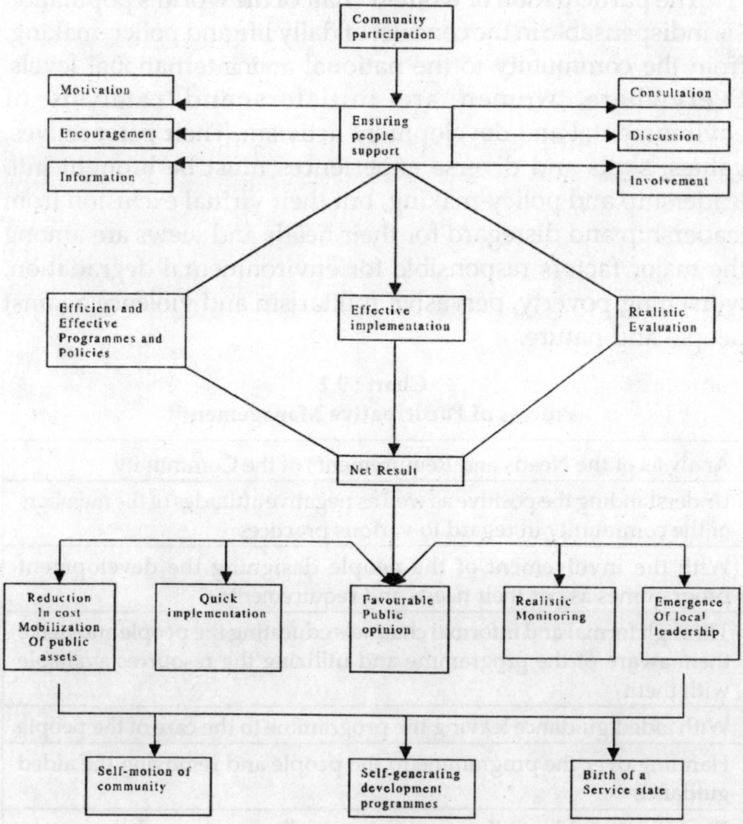

development workers must provide their seasoned guidance to sort out such problems, which may look petty in the initial stages but can lead to disastrous results in the long-run. They should be alive to this fact and must intervene at the opportune moment. If we follow this strategy, it is supposed that the sustainable development programmes may take deep roots in the soil of the community and may continue till there is a need for such programme by the community even for the future generations.

EMPOWERING WOMEN

The participation of women - half of the world's population - is indispensable in the conduct of daily life and policy-making, from the community to the national and international levels. Everywhere, women are initiators and catalysts of environmental and development activism. Their perspectives, values, skills and diverse experiences must be brought into leadership and policy-making, but their virtual exclusion from leadership and disregard for their needs and views are among the major factors responsible for environmental degradation, worsening poverty, pervasive militarism and violence against people and nature.

Chart : 9.2

Process of Participative Management[18]

Analysis of the Needs and Requirements of the Community
Understanding the positive as well as negative attitudes of the members of the community in regard to various practices
With the involvement of the people designing the development programmes as per their needs and requirements
Through formal and informal channels educating the people and make them aware of the programme and utilizing the resources available with them
With aided guidance leaving the programme to the care of the people
Handing over the programme to the people and removing the aided guidance
To sort out problems, if any, occasional follow up is needed
Birth of community managed sustainable development programmes
Birth of a service state

NGOs and governments should comply with requirements of the Nairobi Forward Looking Strategies for the Advancement of Women (approved by the UN General Assembly in 1985), and should ratify and implement the UN Convention on Elimination of All Forms of Discrimination Against Women, the UN Convention of Human Rights, and the UN Convention on the Rights of the Child. People's participation devoid of women power will always remain meaningless and ineffective.

Raising the social, economic, political and health status of women requires policies and actions to assure them of equal access to education, information, fair wages and safe working conditions, inheritance-rights, credit, appropriate technology, environment friendly consumer products and health care. Discrimination on the basis of sex, race, ethnicity, religion, differential ability, and sexual orientation must be opposed; and society should condemn domestic and sexual violence.

No sane person can accept any attempt to deprive women of reproductive freedom, the knowledge required to exercise that freedom, the provision of comprehensive reproductive health care and family planning, including the right to pre-natal care, safe and legal voluntary contraceptives and abortion, sex-education and information. There must be recognition of the existence of a global cancer epidemic which demands the removal from the environment of carcinogenic substances which have particularly adverse effects on women and children, the systematic altering and education of all to the danger of AIDS, and the enhancement of efforts to eliminate occupational hazards in factories, offices and on farms.

During the past twenty years people in various region of India have formed non-violent action movements to protect their environment, their livelihood, and their ways of life. These environmental movements have emerged from the Himalayan regions of Uttar Pradesh (Now in Uttaranchal) to the tropical forests of Kerala and from Gujrat to Tripura in response to projects that threaten to dislocate people and to affect their basic human rights to land, water and ecological stability of life-support systems. They share certain features, such as democratic values and decentralized decision making, with social movements operating in India. The environmental movements are slowly progressing toward defining a model of development to replace the current resource intensive one that has created severe ecological instability (Center for Science and Environment 1982, 190).

Even with limited resources the environmental movements have initiated a new political struggle for safeguarding the interests of the poor and the marginalized, among whom are women, tribal groups, and peasants.

In addition, there are local movements against deforestation, water-logging, salinisation, and desertification in the command areas of dams on the Rapti, Kuano and the terrain of the Gorakhpur district.

Therefore, concept of sustainable development stresses that development must be participatory and involve local people in decision-making process that affect their lives. The people, for whom the development programmes, are meant, must take part in planning and execution for the practical reason, that development efforts which do not, involve local people/groups often fail. In this context, people's/community participation in the development process becomes essential.

NOTE

1. P.S. Ramakrishnan, *op. cit.*
2. FAO, *People's Participation in Development*, Case Studies by Participants of Asian Regional Workshop, volume, II Rome 1973.
3. FAO, *Fifth International Action for Development/FFHC Conference, Rome*, 1971.
4. K.Finsterbusch and W. A.Van Wicklin III, "The Contribution of Beneficiary Participation to Development Project Effectiveness", in *Public Administration and Development*, New York, John Wiley, 1987.
5. A.D. Jedlica, *Organization for Rural Development : Risk Taking and Appropriate Technology*, New York, Praeger, 1977. Also see C. Bryant and L.G. White, *Managing Development in the Third World*, Westview, Boulder Co., 1982.
6. G.Honadle and R.Klaus, (Eds) *International Development Administration-Implementation Analysis for Development Projects*, New York, Praeger, 1978.
7. P.Oakely, *et.al.*, *Projects with People : The Practice of Participation in Rural Development*, Geneva, ILO, 1991.
8. H.J.Eysenck, *et.al.*, *Encyclopaedia of Psychology*, Vol.2, London, Search Press, 1972.
9. J.M.Cohen and N.T.Uphoff, "Participation's Pace in Rural

Development : Seeking Clarity Through Specificity, *World Development"*, Vol. 8, No. 30, 1980, pp. 213-36.

10. S. Paul, "Community Participation in Development Projects : The World Bank Experience", in *Readings in Community Participations, Washington,* D.C. the World Bank, 1987.
11. Uma Lele, *The Design for Rural Development; Lessons-for Africa,* Baltimore, John Hopkins, 1975.
12. Quoted in Cohen and Uphoff, *op. cit.*
13. ACC Task Force on Rural Development, *Report of the Third Meeting of the Working Group on Programme Harmonisation,* Rome, 1978.
14. Santhanam M.L, Community Participation for Sustainable Development, *The Indian Journal of Public Administration,* July-September-1993, Vol. XXXIX, No. 3.
15. D.D. Gow and J. VanSant, "Beyond the Rhetoric of Rural Development Participation: How Can It Be Done", *World Development,* Vol. II, 1983.
16. YASHADA, "Recommendations of YASHDA Workshop on Sustainable Development" held between February 18-19, 1991, Pune, Yashwantrao Chavan Academy of Development Administration, 1991.
17. Mishra S.N, People's Participation for Sustainable Development, *The Indian Journal of Public Administration,* July-September-1993, Vol. XXXIX, No. 3.
18. *ibid.*

10

Human Development and Government Expenditure

Over the past decade, the concept of sustainability has appeared with increasing frequency in the development literature. While sustainability has been applied to a wide range of social and economic issues, including rural development,[1] irrigation,[2] housing and urban development,[3] education[4] and health;[5] most of the literature has focussed on the impacts of economic development policies and demographic changes on the sustainability of the natural environment, and on the short- and long-term impacts of environmental degradation on the quality of human life.[6]

Long-term sustainability will be determined by the extent to which a project contains in-built administrative, financial and organizational mechanisms to ensure continued operation. Bamberger and Cheema suggest that sustainability of social sector programmes should be assessed in terms of four broad groups of indicators :

1. ***Continued Delivery of Services and Production of Benefits :*** This includes the volume and stability of benefits, efficiency of service delivery, quality of services and benefits, beneficiary satisfaction and the distribution of benefits among different social and economic groups.
2. *Maintenance of Physical Infrastructure :* This includes the condition of physical infrastructure, plant and

equipment; adequacy of maintenance procedures; efficiency of cost-recovery; adequacy of operating budget; and beneficiary involvement in maintenance.

3. *Development of a Long-term Institutional Capacity for Managing the Project* : This includes assessing the capacity of the principle operating agencies, stability of staff and budget of operating agencies, adequacy of inter-agency coordination, adequacy of coordination with community organization and beneficiary groups, and flexibility and capacity to adapt the project to changing circumstances.
4. *Continued Support from Key Stakeholders* : This involves assessment of the stability and strength of support from international agencies, national government, provincial and local governments, community organizations, and the ability of the project to avoid becoming politically controversial.[7]

Therefore, the above-mentioned parameters should have also been implemented in Gorakhpur district for sustainable planning and development. All this requires meaningful co-ordination between the government and NGOs, human development of the area and adequate funding. There has not

Table 10.1 : District Urban Development Agency, Gorakhpur Rashtriya Malin Basti Sudhar Karyakram Year 2001-02

Sl. No.	*Head*	*Detail of work*	*Total amount of expenditure*
1.	Book Bank	8	400000.00
2.	Training	-	481350.00
3.	Camp for Mass Awareness Programme	-	168000.00
4.	Centre for the old-aged people	2	194000.00
5.	Information & education	-	44397.00
6.	Plantation	-	71400.00
7.	Catering resources	-	351107.00
8.	Income expenditure from CDS fund	-	801250.00
9.	Balance payment of the last year	-	395555.00
		Total	2907062.00

Source : NSDP & Swarn Jayanti Urban Employment Scheme, 2001-02.

been dearth of financial allocations in the district as in the last financial year 2001-02, District Urban Development Agency (DUDA), Gorakhpur have spent allocated fund on the following heads :

What we can notice is that lack of investment in the health sector, as can be seen from the above table, has not given priority to public health and hygiene. Without healthy population, healthy environment is simply not possible. Not a single rupee has been spent on medical and health. Most of the other heads of expenditure are severely exacerbated by lack of attention to the sustainability of existing services. In a number of these areas, there is a potential for producing very significant reductions in mortality or serious illness at a very modest cost, if along with sufficient funding effective programmes are launched and proper monitoring is done through community empowerment. Only then sustainable development can be achieved in the district.

Investment Problems, Affecting the Continued Delivery of Services and the Production of Benefits

Programmes provide different services and benefits, such as shelter, water, sanitation, maternal and childcare, primary health services, credit, technical assistance and fertilizer. In each case, key sustainability indicators will include the continued provision of these services in the required volume and quality,

Table 10.2 : Graphic illustration

Short Term Action Plan of Gorakhpur (2000-01)	
Project cost	Rs. 2064.71 lakh
No. of Projects	352
Financial Source	Allocation of Fund
Development Authority	Rs. 510.47 lakh
Nagar Nigam	Rs. 651.67 lakh
Jal Nigam	Rs. 456.50 lakh
U.P. Housing and Development Board	Rs. 046.80 lakh
PWD	Rs. 300.00 lakh
Urban Development Agency	Rs. 099.27 lakh
Total	Rs. 2064.71 lakh

SHORT TERM PLANS

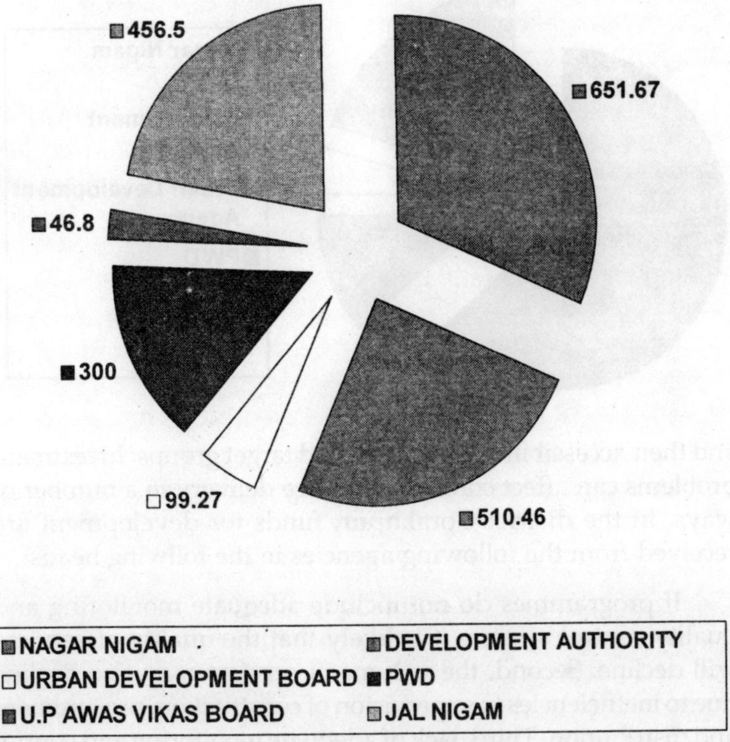

Table 10.3 : Graphic Illustration

Long term Action of Plan of Gorakhpur (2000-01)	
Project Cost	Rs. 10833.86 Lakh
No. of Projects	280
Financial Sources	Allocation of Funds
Development Authority	Rs. 1307.75 lakh
Nagar Nigam	Rs. 0917.39 lakh
Jal Nigam	Rs. 6868.00 lakh
U.P. Housing and Development Board	—
PWD	Rs. 1242.00 lakh
Urban Development Agency	Rs. 0480.72 lakh
Total	Rs. 10833.86 lakh

Long Term Plan (Details of Financial Sources)

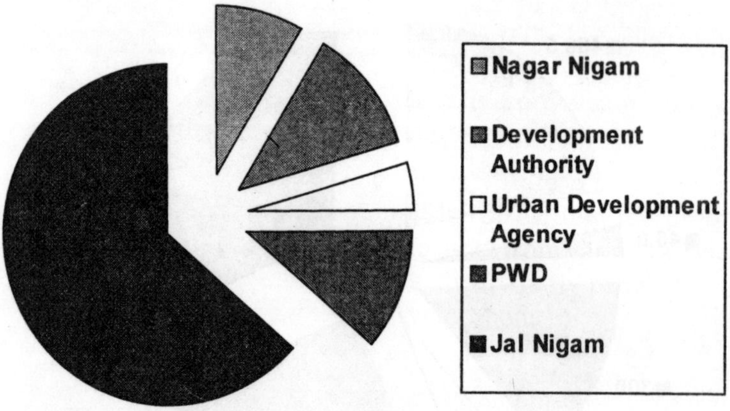

and their accessibility to the intended target groups. Investment problems can affect continued service delivery in a number of ways. In the district Gorakhpur, funds for development are received from the following agencies in the follwing heads:

If programmes do not include adequate monitoring and quality control system, it is likely that the quality of services will decline. Second, the volume of services may also decline due to inefficiencies in supervision of construction, procurement and distribution. Third, lack of adequate monitoring and control can often result in the poorest and most vulnerable groups being gradually excluded from project benefits, as evident from the Pulse Polio Immunization Programme in the district Gorakhpur. During the implementation phase, most projects are closely monitored by the government and donor agencies and consequently procedures designed to protect the poor will be carefully enforced. However, once external attention is diverted to the initiation of other new projects, selection criteria may be relaxed and benefits may be gradually diverted towards more influential groups. For this purpose, DUDA had decided to launch a training programme in the financial year 2001-02, in order to ensure that benefits of various schemes launched for human development, self-employment, poverty alleviation, literacy programme etc. must reach the target. A tabulated version of these programmes and total expenditure incurred can be shown as such :

Table 10.4 : District Urban Development Agency, Gorakhpur Urban Worker's Employment Programme (U.W.E.P.)

Covered Population	575979
Total Financial Target	Rs. 15.35 Lacs
Total Human Days	12531

Table 10.5 : Financial & Physical Targets as per Local Bodies

Sl. No.	*Local Bodies*	*Population*	*Financial Target (Rs. In Lacs)*	*Physical Target*
1.	Pipraich	12523	273344.09	2220
2.	P.P.Ganj	8337	182005.57	1516
3.	Munderwa Bazar	9898	216066.59	1800
4.	Bansgaon	13448	293527.59	2446
5.	Barhalganj	15799	344825.35	2723
6.	Gola	10318	225230.90	1826
	Total	70323	1535000.00	12531

Source : NSDP & Swarn Jayanti Urban Employment Scheme, 2001-02.

Government Expenditure in Health and Sustainable Development

The primary health care system in Gorakhpur has achieved an impressive physical presence in a relatively short period of time and at low levels of investment. The results have been particularly dramatic in the reduction of elimination of malaria and major diseases such as cholera, smallpox, tuberculosis and other respiratory diseases.[8] However, despite these impressive improvements, the health status of the district population remains very low. For example, malnutrition levels, as measured in terms of stunting and wasting were significantly higher than the averages for all of the country. Hospital beds per 1000 population is less than half the average of the state, loss due to communicable and maternal and pre-natal diseases is almost 50 per cent above the Indian average. One reason for the low health status is that due to a number of financial, administrative, infrastructural, and management issues, the health sector had remained neglected. The extensive coverage of other sectors has not been accompanied by the health sector, especially by the rural and urban poor; largely financial and other resources were allocated on more "productive" areas. Thus, many programmes have not been sustained.[9]

Poor quality of construction and maintenance of infrastructure, including offices and staff quarters reduces the efficient distribution of staff and utilization of services. This is often exacerbated by pressure to inaugurate facilities before they are completed. The situation is worsened by unfilled staff position particularly at the medical officer and supervisory levels; and by poor training and preparation of many staff members. For example, a 1988 study of Multiple Purpose Female Health Workers found that 50 percent did not know how to prepare oral dehydration salts or the length of time to sterilize a syringe between injections.[10]

As a result of these and other problems, the rate of utilization of many services by the poor is very low. A 1998 study by Integrated Child Development Scheme (ICDS) and SIFPSA found that in Basgaon, Gola, Sahajanwa, and Sadar Tahsil of the district, only one-third of rural men and 10 percent of rural women used Primary Health Care Services.

Priority is given to the programmes for which quantitative targets have been established. Consequently, it has been estimated that more than half of health workers' activities are directed to Family Planning activities in order to meet 'centrally defined' targets. Other equally important programmes, for which targets are more difficult to define and monitor, frequently receive much lower attention from health workers.

The health care system aspires to provide a broad spectrum of services irrespective of the epidemological and other characteristics of particular areas. At its current level of development, system cannot deliver all these services at acceptable levels of quality.

A key determinant of the long-term sustainability of most social programmes is the provision of adequate financial resources to cover operation and maintenance. The provision of adequate financial resources for project maintenance is particularly difficult in a country as large and diverse as India, but almost all developing countries have similar problems in providing adequate funds to cover operations and maintenance. Most government and donor agencies have a relatively short time horizon and prefer to invest in new infrastructure (which has a politically higher pay-off) rather allocating resources to maintain existing infrastructure. As it is evident from the table:10.6.

While the maintenance of most physical infrastructure, equipment and vehicles and the replenishment of supplies and materials is technically straightforward, and the maintenance requirements are well-known; most governments are reluctant to provide the necessary resources on a regular basis over the life of projects. It has proved extremely difficult for municipalities and project management to generate sufficient revenue to cover these costs. Often the simplest and most equitable way to generate operating revenues would be to charge beneficiaries for the health, education, water or other services they consume. However, there has frequently been strong political opposition to the introduction of user charges—even when many consumers are already paying higher charges for poor quality services obtained through the informal sector.

Inter-agency Coordination

The successful operation of many social sector programmes requires an effective and sustained cooperation between a number of different agencies—each of which has its own priorities and constraints. An IRD project may require active participation of the ministries or agencies responsible for agriculture, health, public works, forestry and water resources. If small business development programmes are involved, it may also be necessary to coordinate with financial and marketing agencies. An urban development project may also involve a similarly broad range of agencies. Unfortunately it has not been

Table 10.6 (a) : Work Plan - 2000-2001
(Financial Arrangement as per agencies)

Short-term programme (Rs. In lakh)

	Institute	*Source of finance*	*Category of Finance*			*Vacant Availability*			*Total*
			Grant	*Debt*	*Total*	*Availability*	*To be available*	*Total*	
1	2	3	4	5	6	7	8	9	10
1	Gorakhpur Development Authority	State Finance Commission	-	-	-	-	-	-	-
		10th Finance Plan	-	-	-	-	-	-	-
		Self Finance	-	-	-	50.00	253.35	303.35	303.35
		Establishment Funds	-	-	-	46.34	80.78	127.12	127.12
		HUDCO	-	5.00	5.00	5.00	-	5.00	5.00
		Others	75.00	-	-	25.00	50.00	75.00	75.00
		Total				126.34	384.13	510.47	510.47

Table 10.6(b) : Work Plan - 2000-2001
(Financial Arrangement as per agencies)

Short-term programme (Rs. In lakh)

	Institute	*Source of finance*	*Category of Finance*			*Vacant Availability*			*Total*
			Grant	*Debt*	*Total*	*Availability*	*To be available*	*Total*	
1	2	3	4	5	6	7	8	9	10
1	Nagar Nigam Gorakhpur	State Finance Commission	168.07	-	168.07	168.07	-	168.07	168.07
		10th Finance Plan	169.56	-	169.56	169.56	-	169.56	169.56
		Self Finance	-	-	-	-	-	-	-
		Establishment Funds	72.06	-	72.06	51.00	21.06	72.06	72.06
		HUDCO	-	-	-	-	-	-	-
		Others	241.98	-	241.98	-	-		241.98
		Total	651.67		651.67				651.67

the case in Gorakhpur. Most of the Government agencies, offices and other institutions working in the field of human development and sustainable development are pathetically coordinated. Either they suffer from financial crunch or affected by changing attitude with the change in government. In the last ten years, Uttar Pradesh has seen dramatic changes in state politics. It has badly affected government expenditure and human development in Uttar Pradesh in general, and Gorakhpur in particular, which has seen even seven ministers at a time in the Council of Ministers during the government of Shri Rajnath Singh, when he was the chief minister in the year 2000-2001. Despite such a huge representation of the district in the Council of Ministers nothing substantial could be achieved on the front of sustainable development. At present also there are several ministers in the State Government belonging from different constituencies of Gorakhpur, still there is no or a very negligible expenditure in the field of human and sustainable development.

In many parts of the country, however, the problem of human and sustainable development is far more difficult, as seen for example, in the terrain and forest region of the Gorakhpur district. Land and water resources are meager. Population pressure is rising. Commercialization in many cases, tend to lead short-term exploitation of resources, and costs of soil degradation and water erosion are not fully reflected in market prices. Traditional, socio-economic systems working at low levels of equilibrium are collapsing and newer methods of sustainable development are not jelling. The Government of Uttar Pradesh has not provided sufficient funds to tackle such problems. Government expenditure in Gorakhpur district on sustainable development, maintenance of ecological balance and human development is almost negligible.

There are innumerable schemes going on land development, rural employment, health and hygiene, child and mother nutrition, poverty alleviation etc. under Jawahar Rozgar Yojna, Indira Gandhi Rozgar Yojna, Integrated Child Development Scheme, Anganbari and Balwari Programmes etc.

in the district Gorakhpur. Huge amount has been spent for this purpose in the last one decade. At the same time there are grants given to local body's representatives, members of parliament and members of state legislative assembly through Sansad and Vidhayak Nidhi. But all these funds have been grossly misutilised. There are reports that government expenditures are pocketed by a coterie of politicians, beauraucrats and other middlemen.

A list of expenditure can be examined in a tabulated form related to programmes of human and sustainable development in Gorakhpur district, given in table 10.7.

New Management Strategy

In the Eighth Plan, heads of expenditure have been decided which could be related to sustainable development in the district Gorakhpur. These programmes were further continued in the Ninth Plan period also.

District agricultural planning and policy has to meet a number of challenges in the years ahead. After a distinct deceleration of output growth and investment efforts in the mid-eighties, the end of the Seventh Plan again saw greater priority being given to the sector both in terms of resource allocation and policy focus. These priorities have to be maintained since the sector has razor's edge quality in relation to India's development for development objectives. A number of feasible investment strategies have been developed for alternative agro/economic people. These have now to be implemented. The investment packages for different region are different and the development strategy has to provide for this differential focus. The land and water development strategies will require a multipronged approach using alternative management strategies most suited to each region. Traditional water harvesting structure, canal improvement systems and improved planning procedures have been established, but need to be implemented, so is the case study of Gorakhpur district proves it.

Table 10.7 : Action Plan 2000-01
Long Term Schemes

Sr. No.	*Name of the work*	*Development Authority*		*Nagar Nigam*		*Jal Sansthan*		*Jal Nigam*		*Awas and Vikash Board*		*Public Works*		*City Development Agency*	
										Project Number	*Amount*				
1	2	3	4	5	6	7	8	9	10	11	12	13	14	15	16
1.	Sewage	-	-	11	36.65	1	4000	-	-	-	-	-	-	-	-
2.	Water supply	2	67.5	20	47.24	3	-	-	-	-	-	-	-	12	15.02
3.	Drainage	3	62.45	-	-	-	-	-	-	-	-	-	-	12	45
4.	Road	2	169.6	123	664.66	-	-	-	-	-	-	1	1242	12	50
5.	Solid Waste Management	-	-	-	-	-	-	-	-	-	-	-	-	20	91.6
6.	Park Beautification	8	150	-	-	-	-	-	-	-	-	-	-	-	-
7.	Removing enchrochment	-	-	-	-	-	-	-	-	-	-	-	-	-	-
8.	Shelter & Others	2	775	-	-	-	-	-	-	-	-	-	-	1	75
9.	Computerization	1	2	6	40.75	-	-	-	-	-	-	-	-	-	-
10.	Miscellaneous	-	-	-	-	-	-	-	-	-	-	-	-	23	56.1
11.	Electrification	4	81.2	2	128.09	-	-	-	-	-	-	-	-	12	48
12.	Date of work ending	-	Mar-02	-	-	-	-	-	-	-	-	-	-	-	-
	Total	22	1307.75	162	917.39	4	6886	-	-	-	-	1	1242	92	480.72

NOTE

1. George Honadle and Jerry VanSant, *Implementation for Sustainability Lesson from Integrated Rural Development*, Hartford (Connecticut), Kumarian Press, 1985.
2. Norman Uphoff, *Improving International Irrigation Management with Farmer Participation: Getting the Process Right*, Westview Press 1986; and Norman Uphoff, (EDI Course Note No. 076/005, March 1989), Washington D.C., The World Bank, 1989.
3. Shabbir Cheema "Sustainable Development : Lesson of Experience" in Michael Bamberger and Shabbir Cheema, *Case Studies of Project Sustainability: Implications for Policy and Operation from Asian Experience* (EDI Seminar Series), Washington, World Bank,1990.
4. World Bank, *Annual Review of Evaluation Results*, Washington D.C. Operations Evaluation Department, The World Bank, 1989; and The World Bank India: *Poverty, Employment and Social Services*, (World Bank Country Study), Washington D.C., The World Bank 1989.
5. Michael Bamberger, "The Sustainability of Social Sector Programme: Lesson from the Health Sector", *Sustainability Development*, 1993.
6. World Bank, *World Development Report 1992: Development and the Environment*, Washington D.C., The World Bank, 1992 and World Resources Institute, *The State of the World*, Washington D.C. World Resources Institute, 1993.
7. Bamberger and Cheema, *op.cit.*
8. B.B.L Sharma, Y.P. Gupta and S. Bhatnagar, "Sustaining Primary Health Care" paper presented at the RIPA/EDI Seminar on "Sustainability of Social Sector Programme in India" in April 1992, at Jaipur.
9. World Bank, *Annual Review of Evaluation Results, op.cit.*
10. World Bank, *India: Poverty, Employment and Social Services, op.cit.*

11
Conclusion and Suggestions

After Independence, as an essential safeguard to attain a just and egalitrian social order, India deliberately opted for state-led development and attempted moblisation of peoples' participation in the development process which, however, remained merely a pious wish. Gorakhpur is a classical case of neglecet and apathy. Despite so much of physical and natural resources, the development of the area has remained grim and grave. Being politically fertile, it has been witness to many travails and tragedies of socio-economic fabric. The political leaders have only used the population as merely their vote banks. Populist measures and unpragmatic policy formulation have only added to the already vitiated problem of development and governance. Administrative officers and beuracrats have either acted as their alibies or remained disenchanted. The picture that emerges is that of a potential area in a state of decay and decomposition. The immediate future seems to belong to the doomsayers rather than to the cheermongers. Our public administration suffers from a fatty degeneration of conscience, and the malady seems to be not only persistent but prone to aggravation.

It is in this context that one hears of the crisis of governability in countries like India.[1] For effective and efficient plan formulation related to sustainable development in the district is an open testimony to the fact that in the last two decades we have taken plenty of wrong turns at the cross roads, misused time, taken gold for dross, and dross for gold. What can be concluded is that the crisis in the political, socio-cultural and economic sphere is that they all compromise the sustainability of society let alone its development. We as a citizen are equally responsible for this sad state of affairs. The quality of our public life has reached the nadir. The four costly failures of the government and the people are:

1. Failure to maintain law and order. Peace is a prerequisite for progress and development. The area has always ridden with criminals and anti social elements. The standards of our politicians, policemen and criminals are indistinguishable from one another, making any type of development impossible. We have too much government and too little administration, too many public servants and too little public service, too many controls and too little welfare, too many laws and too little justice.

2. Failure to bring the unbounded economic potential of the area to fruition. The perennial river water systems of the area, the large amount of biodiversity present in the forests, vast human resources etc., all have remained untapped or misutilized.

3. Failure to make human investment- investment in education, family planning, nutrition in public health, in contradistinction to physical investment in factories and plants, that too could not prove purposeful. Gross National Happiness should have been given priority over Gross National Product (GNH>GNP).

4. Failure to provide effective and visionary leadership either by the community or by the beuracracy. When a republic comes to birth, it is the leaders who produce the institutions. Later the institutions produce the leaders to guide and govern the masses and community. The question- where are the leaders of tomorrow at the district level?-can only be answered by another question- where are the Nation Building Institutions which can

produce the leaders of tomorrow? We have failed at all the fronts.

Therefore, what we need today is general awareness and mass mobilization of both the administrative machinery and the common man for propagating and practising concept of sustainable development. Do we have educational institutions in the area which aim at generating excellence and are equipped to produce movers of peoples, mobilizers of opinion-integrated personalities whose minds, hearts and character have been developed in the noble traditions of our invaluable heritage?

Animals can be trained, only human beings can be educated. It requires personal participation; therefore, more and more people from all ranks and file of society should be involved to develop ideas and behaviour of sustainable development which can easily be executed locally. An attempt is therefore being made here to offer a broad outline as to what constitutes sustainable development but also how we can achieve it:[2]

1. It can not be achieved in the short run.
2. It is based on equity and justice.
3. Its approach is balanced and intergrative.
4. It has common goal but different routes.
5. It accepts nature not only as a resource for development but also as the earthly womb for survival and development of humankind, and
6. It is participatory in nature.

Therefore, sustainable development is not the business of the government and private companies alone. It is the business of the people in general. It is a process which has to be initiated at each level of human endeavour and life. It involves individuals, families, communities, corporate bodies, nations and global societies. It has to be a movement, because it involves paradigmatic change which is difficult to bring about unless many people get involved. Democratization of decision-making and decentralization of power and authority should be achieved. Corruption is the greatest solvent of public institutions, poverty posses a far smaller threat. All developmental programmes become redundant or are found to exist only in official papers because of the rampant and unbridled corruption from top to bottom. All are responsible for this malaise. It has become very difficult to formulate programmes and policies which may bring in sustainability in development because of vested interests and no concern for tomorrow. The whole area has become a sick man of Uttar Pradesh. The district needs an administration today, which is strong without being authoritarian and humane without being weak. Therefore, the filth and squalor scattered all around the rural and urban areas of the district cannot be contained by merely slogan mongering or eloquence. It requires effective planning and strong political will along with community participation.

We must establish systems of review of emergent and existing technologies. Our ability to alter the environment has grown much faster than our understandings of the environment. Many negative environmental impacts are the result of unanticipated side effects. New technologies that have emerged in recent decades have been particularly troublesome. We now need systematic controls which insist that such technologies be ecologically sound. There is after all nothing inherently bad about technologies. The task is clearly one of taking stock of where we are and where we would like to go with our technologies and to consider of our destiny, in other words.

We must establish more environmentally cognizant approaches to technology transfer from "developed" to "developing" countries. We live in a diverse world-diverse in biomes, cultures, and development. All human beings obviously have similar needs for food, clothing, and shelter. There are also similar desires for a high standard of living that now constitutes a conflict between what is fair and what is possible. Like it or not, the world cannot stand the spread of energy-intensive, material-intensive cultural practice to place where they do not yet exist.

Technologies and economies must be developed that are ecologically more appropriate and sustainable. The current level of production and consumption in the United States cannot be maintained indefinitely. Our technologies are environmentally too costly. It seems that we must now begin to move to smaller-scale and environmentally more appropriate technologies, as are advocated for developing nations. It is unfortunate that we have for a time put all of our eggs in the fossil fuel basket. And it would be unfortunate if we placed too heavy an emphasis on any one new energy source. We must begin to diversify again-to have technologies and way of doing things that differ from place to place according to differences in local environments.

We need environmental multidisciplinarians. The task before us requires that some of us become trained to deal with multidisciplinary problems. The optimal format for such training would probably consist of a solid grounding in major discipline, but with training in a number of others as well. As harmless as this might sound, it does go against the grain of the education and occupational trends of the "age of specialization."

Recycling must become a way of life. The problem with material resources is one of keeping materials in forms in which they are usable. There has been some movement toward recycling in industry, but what we have accomplished so far really amounts to just a tentative first step. If we can make recycling an important part of all of the things we do in society, we will have solved several kinds of problems.

We must develop stragtegies to encourage source reduction of pollution. Many of our pollution problems are the result of systems that exploit resources and add nothing to our standard of living. Much of our present packaging falls into this category. An example is jars packaged in boxes that are promptly tossed away, adding to the solid waste problem. Disposable containers are another example.

We must continue to internalize environmental costs into the cost of doing business. We must make it more expensive to pollute. If the environmental costs are included in the price of the product, consumers are more likely to buy the less expensive, more environmentally compatible products. Ways must be found to internalize environmental costs along with other production costs.

Charles E. Kupchella and Marget C. Hyland[3] rightly say that, "Where conflicts exist between individual freedom and the quality of the human environment, individual freedom must give way. ....we have described our traditional reverence for individual freedom and how this sometimes obstructs intelligent stewardship of the environment. We have implied that (despite its not being a popular notion) government might actually have to assume a larger role in human affairs if the environment is to be looked after properly. If there are critics who would brand this attitude as an environmental form of socialism, so be it. We recognize that societies are made up of individuals and that, in general, as an individual goes, so the society. This is true only up to a point, however, and we believe that many of the environmental problems we face are direct consequences of our having carried this ideal past the point of optimization."

Ecological paradigms must become the basis of all ethical systems. Our religions must place more emphasis on environmental stewardship. The basis of our system of laws must change from its emphasis on property ownership and possession of an emphasis on property stewardship and responsibility. Under such a system, destruction of "one's own" land so that future generations are deprived of it, should be considered an act of theft. The decision makers in a corporation that pollutes groundwater resource should be held legally accountable and be appropriately penalized.

A few suggestions that can be mooted for sustainable development are[4]:

1. The carrying capacity of an eco-system should be enhanced through adoption of cleaner technology and efficient management package;
2. Inculcate scientific temper not merely among the scientists, engineers and technologists, but even more

so among the decision makers-politicians, bureaucrats and technocrats;

3. Reorient the working of various ministries depending heavily upon crucial technology inputs through induction of professionals fully conversant with the subject at all levels of decision-making;
4. Revise personnel policy to facilitate result-oriented recruitment and promotion by discarding the present time-scale promotion of personnel from various services so that beuraucrats take interest in programmes concerning sustainable development.
5. All positions of the level of Director and above to be filled on the basis of competence and merit only through open competition;
6. Adoption of a system of contractual hiring of personnel on the basis of performance by fixing the responsibility and accountability;
7. Facilitate easy mobility of professionals among the universities, laboratories, industry, and the ministries on the basis of goal-oriented programmes and contractual appointments;
8. Reorient the educational system to encourage creativity by imparting skills on the basis of aptitude and interests rather than mere acquisition of degrees;
9. Undertake carrying capacity based planning through harmonization of sectoral policies and restructing of the organizational set-up at the state and Central level by adopting water shed as a planning unit;
10. Shift away from a regime of subsidies to meaningful productivity and quality improvement through pursuit of excellence at all levels;
11. Ensure transparency in decision making with 'Right to Know' made operational so that the public can, if needed, know how and why a decision has been arrived at; and

12. Enshrine pursuit of excellence through suitable incentives and actively discourage sycophancy through disincetives.

In a final sense sustainable economic growth, removal of poverty, hunger and malnutrition, are not merely a function of economic planning or techonological innovations but stems from a moral force which accepts no hunger, nor the widening disparities between the rich and poor nations. The success of the programmes related to sustainable development will be measured not only by raising more funds, organizing more conventions and conferences, or establishing more environment institutions but essentially by the inner change and the practice of the three reconciliations mentioned earlier, by the emphasis on *ahimsa, antyodaya and austerity* and by following the *Dhamma* of Buddha. Finally, environmental concerns are moral and ethical and until we recoginse that there will be only talk and no action.

NOTE

1. Atul Kohli, *Democracy and Discontent : India's Growing Crisis of Governability*, Cambridge, Oxford University Press, 1991.
2. R.P.Mishra, Sustainable Development: The Ecological Perspective,' *The Indian Journal of Public Administration*, July-September, 1993.
3. Charles E. Kupchell & Margert C. Hyland: *Environmental Science: Living Within The System of Nature*, Second Edition, 1989.
4. S.Maudgal, Organisationl and Administrative Restructuring for Sustainable Development, *The Indian Journal of Public Administration*, July-September, 1993.

Bibiliography

Alagh, Y.K., (1986), "The Agricultural sub-model", in *Studies on the Structure of the Indian Economy and Planning for Development*, PPD, Planning Commission, New Delhi, Ch. 2.

Alagh, Y.K., (1986), "Regional Dimension of Indian Agriculture", in Papanek, Gustav, and R.B. Lucas (eds.) *The Indian Economy*, Oxford University Press, Oxford, pp. 29-52.

Alagh, Y.K., (1988), "Guidelines for Agro-Climatic Planning – a Draft for Discussion", *Journal of Land Development*, Revenue Board, Lucknow, Vol. I, No. I.

Alagh, Y.K., *"Agro-Climatic Regional Planning : An Overview"*, Planning Commission, New Delhi.

Alagh, Y.K., S.P. Kashyap and G.U.S.N. Morthy, (1984). "Policy Modelling for Planning in India", in S.K. Choen and Erik Thorbeik (eds.) *The Modelling of Socio-Economic Process*, Gower, Aldershot, pp. 5887.

Alagh, Y.K., (1983), *"Report of the Agricultural Price Commission for the Rabi Season"*, Manager of Publications, New Delhi.

Alagh, Y.K. and G.S. Bhalla, (1988), *"Performance of Indian Agriculture : A District-wise Study"*, Sterling, Delhi.

Bhalla, G.S. and D.S. Tyap, (1989), *"Pattern in Indian Agricultural Development"*, ISID, Delhi.

Bryant, C. and L.G. White, (1984), *"Managing Rural Development with Small Farmer Participation"*, West Hartford, Kumarian Press.

Buch, M.N., (1992), "Population and Environment" in V. Gowariker (ed.) *Science, Population and Development*, Unmesh Communication, Pune.

Chopra, R. and D. Sen, (1992), "Natural Bounty and Artificial Scarcity In Water Resources", in V. Gowariker (ed.) *Science, Population and Development*, Unmesh Communications, Pune.

Chopra, Kanchana, K.K. Gopal and M.N. Murty, (1989), *"Participatory Development: People and Common Property Resources"*, Sage, London.

CSE, (1987), *"The Wrath of Nature - The Impact of Environmental Destruction on Floods and Droughts"*, Center for Science and Environment, New Delhi.

Dandekar, V.M. and F.K. Wadia, (1989), "Development of Agricultural Administration In India", *Journal of Indian School of Political Economy*, pp. 1-28.

Dev, S. Mahendra, (1985), "Directions of Change in Performance of All Crops in Indian Agriculture in the Late Seventies", *Economic and Political Weekly*.

Gadgil, Madhav, (1994), "Biodiversity : Reckoning With Life", *Survey of the Environment*, The Hindu, pp. 23-30.

GOI, Planning Commission, (1981), *"A Technical Note on the Sixth Plan of India"*, Planning Commission, Government of India, New Delhi.

GOI, Planning Commission, (1985), *"Report of the Agricultural Prices Commission for Rabi Season"*, Manager of Publications, New Delhi.

GOI, Planning Commission. 1987. *"Mid-Term Appraisal of the Seventh Five Year Plan"*, Planning Commission, Government of India, New Delhi.

GOI, Planning Commission, (1987), *"Involvement of People in the Planning Process"*, Manager of Publications, Delhi.

GOI, Planning Commission, (1988), *"Framework Action Plan for Foodgrain Production"*, Planning Commission, Government of India, New Delhi.

GOI, Planning Commission, (1988), *"Framework Action Plan for Foodgrain Production"*, Planning Commission, Government of India, New Delhi.

GOI, Planning Commission, (1988), *"Guidelines for Planning at the Agro-Climatic Regional Level"*, Planning Commission, Government of India, New Delhi.

GOI, Planning Commission, (1989), *"Studies on the Structure of the Indian Economy and Planning for Development"*, Planning Commission, Government of India, New Delhi.

GOI, Planning Commission, (1989), *"Agro-Climatic Regional Planning: An Overview"*, Planning Commission, Government of India, New Delhi.

GOI, Planning Commission, Ministry of Finance. *"Economic Survey of 1989-90"*, Manager of Publications, New Delhi.

GOI, (1992), *"The National Conservation Strategy and Policy Statement on Environment and Development"*, Government of India.

Kunda, A. and M. Raza, (1982), *"Indian Economy : The Regional Dimension"*, Spectrum, Delhi.

Mathew, George, (1995), "Asia: Emerging Socio-Economic and Political Issues", *The other Side-A Journal of Socialist Thought and Action*, No. 134, pp. 7-12.

NSS, *Report No. 5* (Lahiri D.B.), *Report No. 27* and *Report No. 70* (Murti M.N.), *Report No.125* (Roy A.S. and Bhatacherya A.) and *Technical Aspects of Design: A Collection of Five Years Papers,* Indian Statistical Institute, New Delhi.

OED, (1986), *"Sustainability of Projects: Review of Experiences in Fertiliser Subsector"*, Operations Evaluation Department, Washington D. C., The World Bank.

OED, (1989), *"Impact Evaluation Report : Indonesia – Sustainability of the First and Second Education Project"*, Operations Evaluation Department, Washington D. C., The World Bank.

Rao, Mohan J., (1995), "Whither India's Environment", *Economic and Political Weekly*, XXX (13), pp. 677-686.

Sahu, K.C., (1994), "Power Plant Pollution – The Cost of Coal Combustion", *Survey of Environment*, The Hindu, pp. 437-52.

Sharma, S.L., (1992), "Social Action Group As Harbingers of Silent Revolution",*Economic and Political Weekly*, Vol. 27, No. 47.

Sharma, S.L., (1993), "Managing Environment Crisis-Regulatory vis-a-vis Participatory Approach" in J.L. Rastogi and B. Chand (eds.) *Management and Business Education*, Rawat Publication, Jaipur.

Sivaramakrishnan, K.C., (1993), *"Managing Urban Environment In India-Towards an Agenda for Action"* (3 Volumes), The Times Research Foundation, Calcutta.

Singh, Daman and Manjul Bajaj, (1988), "Usar Soils in U.P", *Economic and Political Weekly*, pp.A.181-187.

Singh, J.P., (1988), *"Evaluation of Centrally sponsored Scheme of Special Rice Production Programme in Eastern Uttar Pradesh"*, Agro-Economic Research Centre, University of Delhi.

SPWD, (1988), *"National Workshop on Small Scale Development: Proceedings"*, Society for Promotion of Waterlands Development, New Delhi.

Santhanam, M.L., (1985), "Strategies of Organising the Poor for Development", *Partnership in Progress* Issue II, Calcutta, Luthem World Services.

WRI, (1992), "*World Resources 1992-93*", New York, World Resources Institute, Oxford University Press.